BRITAIN'S
TOY CAR WARS

BRITAIN'S
TOY CAR WARS

THE WAR OF WHEELS BETWEEN
DINKY, CORGI & MATCHBOX

GILES CHAPMAN

The
History
Press

Frontispiece: Aston Martin DB5 James Bond toy cars being produced at Corgi's Swansea factory. (Alamy)

First published 2021

The History Press
97 St George's Place, Cheltenham,
Gloucestershire, GL50 3QB
www.thehistorypress.co.uk

British Library Cataloguing in Publication Data.
A catalogue record for this book is available from the British Library.

ISBN 978 0 7509 9713 3

Typesetting and origination by The History Press
Printed in Turkey by Imak.

Contents

Introduction

This is the intertwined story of three iconic names in British toy-making. It's the most detailed, the deepest, and the most comprehensive examination of the contrasting stories of Dinky Toys, Matchbox and Corgi Toys, and how they competed intensely with each other for thirty solid years.

These world-famous products were locked in a ferocious battle for the attention of boys and, crucially, their parents and benevolent relatives, whose custom turned their manufacturers into world-renowned businesses.

The name of the game was excellence in miniaturising the machines that boys (and, sorry ladies, but it was predominantly us chaps) fixated upon or dreamed about. Mostly these were cars but the British die-cast phenomenon also extended to trucks, tractors, motorcycles, buses, aircraft, tanks and ships, and later on formed tie-ups with films and TV shows to add new elements of fantasy and inspired design ingenuity.

From their three corners of Britain, Meccano's Dinky Toys, Lesney's various Matchbox series and Mettoy's Corgi Toys took distinctive and individual approaches to what they produced. As they fuelled imaginations and filled up toy boxes the world over, the skill of the three businesses at extracting pocket money from children and providing gift answers to perplexed grown-ups led to stellar corporate growth, profits and, indeed, acclaim for a world-beating British industry.

Yet from highpoints in the mid/late 1960s, all three of these legendary toy-making enterprises endured a rocky road in the turbulent society of the 1970s. The nation's three toy car heroes drove onwards, updating and refreshing their wares to stimulate as much 'play value' as possible. It was even tougher as the decade turned the corner into the 1980s. Tastes, trends, technology and, indeed, teenagers changed rapidly; the complex and carefully crafted manufacturing processes of which each company was so justly proud suddenly became a burden, cumbersome and uncompetitive. If you were passionate about real-life British cars and lorries, then the parallels with the country's declining motor industry of the time were especially poignant.

They were all to stumble and, ultimately, disappear as the digital era dawned. It's like it never happened … except for the fact that these three companies' obsolete products are now collected so avidly that five-figure sums have changed hands for single items originally intended to be mass-produced in their millions.

It's impossible for me to relay this saga without including my own thoughts and interpretations. I've loved Matchbox, Corgi and Dinky vehicles my entire life. Actually, I'd go further. They've had a profound effect on me. I was born with a congenital eye problem called aphakia, rare in children, that meant my lenses were malformed, and so I had no focusing power at all.

Are Toy Cars Still Relevant?

In modern life, with all its digital distractions and instant entertainment, toy cars don't exert the pull they once did. But the fact that they are important to people's lives was recorded in 2015 by research undertaken by pollsters Opinion Matters and funded by carmaker Honda. They found one in five adults admitted to still playing with toy cars and that one in ten wouldn't be embarrassed to do so in the company of other adults. Indeed, 13 per cent said they had received one as a gift in adult life and then, when buying them as gifts for others, a quarter of the survey said they enjoyed the process because of the happy memories it brought back of childhood. A third of these adults said they had hung on to toy cars from their youth, with 10 per cent saying they kept them stored safely away, and a similar number saying they had them on proud display at home.

It's hard to explain how this feels to a person with normal eyesight, but if you try looking the wrong way through binoculars or a camera lens then you might get some idea of how frustrating the distant, uncorrected, unfocused effect is. Back in the mid-1960s, specialists weren't able to operate on me to correct it and in fact chose to bide their time until I was five years old to get some sort of cogent spoken feedback from me about what I could and couldn't see. Then they could assess the compromised state of my developing vision and I could be furnished with glasses.

I do have a vague memory of putting on my first specs. I was strangely reluctant to wear them at the start but they provided a revelation in clarity. Objects were distinct, their edges unbelievably crisp, and I could make out letters in a book rather than from prompt cards with gigantic type sizes. There followed several days of wandering about simply examining things at home and these naturally included my toys.

Now I could see what my toy cars were really like, up close. Peering at them before had been hopeless, of course. One, in particular, captivated me. It was the Matchbox No. 12 Land Rover Safari. Before, I could trace its features with my fingers but not really see them and now I could examine its fully detailed interior and the fine moulded outlines of its doors, radiator grille, even a number plate. Most impressive to me was its roof-mounted cargo, an intricately recreated brown plastic representation of tightly grouped suitcases and containers secured for a long sortie into inhospitable terrain. It was spellbinding to hold this in my hand and then, later, recognise a version of the real thing on the road from the back seat of our family car.

Toy cars became my passion, the only things I wanted for my birthday and the possessions I most coveted (if they were different from my own) of my friends.

I began to accumulate a pretty big collection from the three main British manufacturers. Dinky, Corgi and Matchbox were all stocked by our local toyshop, Sidney W. Dine Ltd of Scartho, just outside Grimsby in Lincolnshire. The latest releases were showcased in a lockable cabinet near the shop door, while shelves behind the counter groaned with models in their tantalising packaging, sporting an abundance of bright yellow and red to snag and mesmerise the young addict. I would take my pocket money in there and spend an hour or more weighing up what to get next, while the proprietor and his assistants in their brown shop coats sighed and tutted.

The vehicles I really desired were almost always beyond my means, but quantity was important and I wanted something new every week. Rather than go without my 'fix' for seven days and save up for a temptingly presented Corgi gift set or an impressive Dinky Toys lorry, I was often tempted by Dine's strategy (the period in question is 1974–78) of offering 'old stock' at reduced prices. And this triggered in me a fascination for the changing style of model design, together with the graphics and typography of the packaging. The old stuff, even then, seemed to me finer, as though the newer products had been demeaned by cheapening, garish paintwork and a general dumbing down of detail. At that time, though, I couldn't begin to guess at why.

Yet I was aware of the great venerability of, in particular, Dinky Toys. My father still had all his from a childhood in the 1940s and early '50s and, although I was allowed to play with them, he removed them stealthily from any situation involving potential damage and kept them very much 'his' rather than 'ours'. Of course, in the mid-1970s, pre-war Dinky Toys were just emerging as 'collectibles' and yet the self-same brand, now with plastic parts and lurid paintwork, was still on sale at our village toyshop. I found it all very intriguing.

My collecting took a quantum leap in my young teenage years. Family friends and relations, aware of my passion, would give me box-loads of old toy cars when lofts or grown-up sons' bedrooms were cleared. But also, now owning a bike and allowed to go where I wanted on it on Saturdays, I picked out more at junk shops and jumble sales, purchasable with a few coppers. Often, with no reference sources, my finds were totally new and strange, such as a delightful Alfa Romeo Spider made by an outfit called Lone Star, or a superb Bentley stamped underneath 'Spot On Models by Tri-ang'. I was beginning to piece together fragments of information about British die-cast toys and their fascinating provenance even though there were no reference sources I knew of to consult. Each second-hand one I got my hands on was quite literally a discovery.

I was frequently left bewildered by the very appearance and aura of the toy vehicles themselves. Almost anything carrying the Corgi Toys stamp, for example, exuded a true car enthusiast's stickling for authentic detail. It was as if the unseen and unknown driving force behind their creation read and absorbed the same car magazines as I did, and chose high-performance subjects to model that were directly intended, uncannily, to pique my own interest. The anomalies were confusing. How come Corgi's Range Rover had contours and wheels that made it look so hopelessly unfaithful to the real thing while the one from Dinky Toys – more normally the brand with the clumsiest hand at miniaturisation, I found – captured this desirable car so brilliantly? Full-size vehicles often had a 'stance', such as the angled, nose-down profile of the Citroën Dyane, which both Corgi and Dinky artfully recreated at small scale. But then there was no other car in the real world to resemble the Mini, and the various iterations produced by Dinky, Corgi and Matchbox represented the same thing seen through wildly different eyes. Britain's most recognisable real-life motor car seemed to warp alarmingly in the hands of the country's three leading toy car makers.

I first attended a 'swapmeet' in about 1981, in Mansfield. Swapmeets had once been the exclusive preserve of old toy-train fanatics, meeting up to trade in locos and layout scenery, but now these specialist collectors' fairs were welcoming in the old Dinky Toys that had begun to demand big money at vintage toy auctions in London. Seeing some of the prices demanded for original boxes, I was glad I'd instinctively kept many of mine.

Shortly thereafter, my interest waned as other teenage leanings took over. Aged 16, admitting to collecting toy cars was never going to sound cool to any girl. However, I had reason to be grateful to my extensive hoard of Corgi, Dinky and Matchbox when years later the time came to buy a flat. Selling them off helped towards the deposit.

Anyway, that was all a long time ago. Now, with so much researched and published about the three rivals, it's finally possible to mix my personal experience of collecting with the incredible stories of these three legends of British toy manufacturing. There have been plenty of reference books on one or other of the brands, most of them tackling the tangled matters of numbering, issue dates, specifications and values. For Corgi, the design genius behind the range of products for three decades managed to compile a reference book of unparalleled completeness long before he took his knowledge to the grave with him. In the case of Dinky and Matchbox, some of the most definitive reference books have depended on the diligent detective work of amateur collectors, whose accumulated discoveries and cataloguing have built up the massive knowledge base that exists today. Certainly, these two manufacturers themselves never made understanding their history easy and mostly treated really avid collectors with bemused indifference.

Then again, all three companies had more pressing issues of survival to wrestle with than worry about toy car-mad and truck-mad foamers. This is a saga – the story of their rise and fall – that runs from the just post-Depression era of the early 1930s, via the 'You've never had it so good' period of the 1950s, through the soaring exuberance of the 1960s and the rocky ride of the 1970s, culminating in Margaret Thatcher's arrival in Downing Street. As British industry was decimated by dreadful industrial relations and devastating economic conditions, Meccano, Lesney and Mettoy faced the onslaught and all collapsed. If you'd been reading the City pages in your daily paper, you might have been aware of their plight and inevitable demise; you might have understood the wider circumstances and the strategic mistakes the companies made. However, if you were merely a keen consumer of their wares, you probably wondered, in bewilderment, what on earth could have gone so terribly wrong.

So in this book we begin at the birth of Britain's craze for the miniature vehicle cast in metal in the early 1930s and retrace every step of this blockbuster product's odyssey through our childhoods to the point where the legendary names behind them effectively ceased to exist. We're going to find out what really happened. We'll be taking this journey with the enormous benefit of hindsight, copious historical facts and information, revealing interviews with those intimately involved, and, if you'll allow me, plenty of personal observations, musings and conclusions from decades of collecting, buying and selling.

Author's Note to this Edition

The first version of this book was published in 2016. The idea for it came to me a year earlier as something different and complementary to all the incredible reference sources already published, many of which are acknowledged in the Bibliography. I was elated when The History Press liked the concept and agreed to publish it ... and amazed that that book itself has now become collectable and quite hard to find. However, rather than just reprint it, I wanted to produce a heavily revised edition expanded and updated with new research, interviews and information. Once again, I'm so delighted The History Press, especially Amy Rigg and Gareth Swain, share my belief that this story strikes a chord with possibly millions of people who were once mad about toy cars, and possibly still are, and who will enjoy this even more probing exploration behind the scenes of Dinky, Corgi and Matchbox! If I am a geek about it all then fine, very happy to be so, and will be even happier if this tale satisfies other grown-ups for whom a part of them is forever lying on the living room floor, toy car in hand, making muffled engine noises ...

1

Die-casting and Modelled Miniatures

Frank Hornby, born on 18 May 1863, never did receive the knighthood he so richly deserved. His brief spell at the heart of the British establishment was confined to a single term as Tory MP for Everton and even that was cut short by his death in 1936.

Yet Hornby built a business empire with an uncanny knack for innovation, marketing and exports. He was one of the most dynamic entrepreneurs of his time – brilliant at turning promising contemporary ideas into products that people wanted. Simultaneously, he created and nurtured the groundbreaking Meccano construction toy and Hornby Trains, both of which became major influences on male British childhood. Then, almost by accident in 1934, Dinky Toys was to join his stable of world-famous brands.

Mr Hornby began his working life as a bookkeeper with a Liverpool meat importer. A respectable family man who worked hard during the day, he was a skilled amateur metalworker and in slack moments when he wasn't occupied with his wife and three children he relaxed by building metal models of bridges and cranes for his sons Roland and Douglas, using skills gained by working briefly for his own father, a model engineer. It was very much the nature of playthings of the time that they were mostly home-made. The British toy 'industry' was in its infancy and the average white-collar employee, with no welfare state as backup, was hard-pressed to meet his bills and pay for a home, never mind buy toys for the family.

Therefore, Frank's 'Eureka moment' was to devise a metal construction toy with the versatility to be assembled into various contraptions that could then be taken apart and the parts used to build something else entirely. He must have been perpetually exhausted, because he poured countless hours in his evenings and weekends into development, but the work was rewarded by a ringing endorsement from Professor Henry Hele-Shaw of Liverpool University. He was fulsome in his praise for what he recognised as a toy that could get boys excited by the engineering possibilities of metal strips and brackets with drilled holes that could be joined together using nuts and bolts.

Prof Hele-Shaw, who was also president of the Society of Model and Experimental Engineers, was happy to commit his opinion to a letter, and Hornby used this as part of his pitch to his employer, David Elliott, to lend him the £5 necessary to take out a patent. Only, Elliott thought the product was such a potential winner that he instead proposed a formal partnership with the earnest new inventor and funded a workshop next door to the meat emporium, where Hornby could concentrate on perfecting his ideas, manufacturing the parts, and packaging up his new product as Mechanics Made Easy.

It was, at first, a tough sell. Wholesalers were often sniffy about the relatively high 7s 6d cost of a set and the fact that it was pseudo-educational mostly left them cold. Only when Mechanics Made Easy built-up models were placed in shop window displays did the penny drop and when, in 1907, the catchy new Meccano name was registered, sales began to move.

And it wasn't just in Britain. Foreign importers couldn't get enough of Meccano (which is easy to pronounce in any language). Hornby was swift and adept at ramping up production of his rapidly growing arsenal of bolt-on components and by 1912 he'd also established Meccano France with his 23-year-old, French-speaking son Roland at its helm.

However, these deft moves seemed small beer next to Hornby's acquisition, in 1914, of an enormous, 216,000sq ft factory at Binns Road, Liverpool. Even rivals grudgingly conceded that it was probably the best-equipped toy factory in the country, and Hornby brought in advisers to help him plan the separate departments and organise really large-scale manufacturing.

After this impressive entrepreneurial splurge, in 1920 Meccano launched its first toy model railway set under the new banner of Hornby Trains. Meccano first offered clockwork motors, imported from Märklin in Germany, in 1912, and in 1915 marketed a train set also thought to have been made by Märklin. The Hornby Trains, though, were made entirely at Binns Road, and were ingeniously constructed along Meccano's familiar nut-and-bolt principles. The sets came with ready-built models, albeit ones that could be taken to pieces because they were assembled from, well, basically Meccano.

Ever the perfectionist, Hornby – whose own name was on this product – was stung by criticism that the trains looked clunky and he ordered that they quickly evolve to become much more lifelike. Hornby soon copied successful German toy firms such as Märklin in adopting the O gauge for his track, setting the scale at about 1:48. The emphasis switched from tinkering with the trains themselves to building an interesting and realistic layout and the company created numerous tinplate and wooden accessories for added authenticity. The first ever set was officially called the Hornby Tinprinted Train Set. This catalogue title unwittingly highlighted one of the limitations of metal toys at the time.

Tinplate strips could be bent and turned to give a decent amount of three-dimensional detail. However, intricate representations of the small parts of a real railway steam locomotive – such as its exterior pipework – had to be printed on to the tin sheet, just as elaborate decorative detail was printed on to tins for things such as biscuits and mustard. To sustain Meccano's pace for innovation, and to feed the detail freaks that youngsters were rapidly becoming, another technique was needed and it lay in the latest developments in die-casting.

The basic die-casting process is simple to understand. It's the forcing of molten metal into a shaped cavity under high pressure. The two parts of the 'die' together form a negative of the object being cast inside and, once the metal has cooled and solidified, it will have adopted every contour of the inner die sections. The process was a refinement of the basic casting process that stretched back through the centuries. Only in 1849 did one J.J. Sturgis successfully lodge a patent for the first manually operated die-casting machine, which was used for casting printer's type in lead, and you can fully understand the impact and prospects this represented to the publishing industry. In 1893, at The World's Columbian Exposition in Chicago (later known as the Chicago World's Fair) the Line-O-Type type-casting machine was announced by its inventor Otto Mergenthaler and within two years this automated apparatus was in operation providing type for newspapers, which it would do right up to the 1970s. The fine detail such equipment allowed was soon harnessed by other manufacturers of consumer goods, everything from door knobs to parts for gramophones and cash registers, which could be churned out in huge quantities to hitherto unknown standards of consistency. When hydraulically assisted machines came on stream from around 1910, manufacturing volumes ballooned.

Many businessmen with new ventures in mind were excited by the possibilities that die-casting offered and one of them was Charles O. Dowst, a Chicago bookkeeper who, in 1879, took his first steps into publishing. Two years later he and his brother Samuel launched *The National Laundry Journal* to capitalise on a fast-growing sector of the US economy.

The Dowsts were, naturally, early adopters of a new Line-O-Type machine, but as it performed its task for the pages of the *Journal*, the brothers realised it could be adapted to other uses. They were soon casting a range of tiny items vital to the laundry industry, including collar buttons, cufflinks and tiny promotional irons. These irons, indeed, were soon part of a range of metal charms that also numbered ships and animals. They were sold to the confectionery trade as cake decorations, free gifts in boxes of popcorn and eventually, and most famously, used as playing pieces in the board game Monopoly.

In 1906, as the prospects were brightening for Frank Hornby's Mechanics Made Easy on the opposite side of the Atlantic, Samuel Dowst's son Theodore S. 'Ted' Dowst joined the Chicago family casting business as a clerk. It may have been his inventiveness that caused the firm to add a model car to its range in 1911, a tiny limousine made of lead with the innovation of rotating wheels. By all accounts, it was an instant hit.

This and other items opened up a new avenue for Dowst Brothers as toy manufacturers. They found a particularly strong-selling line in furniture for dolls' houses, which from around 1921 was sold under the Tootsietoy brand name. It was officially trademarked on 11 March 1924, although few could have known that Tootsie was actually Ted Dowst's affectionate nickname for his daughter Catherine. Even fewer knew that she was the illegitimate result of a liaison between Dowst and a secretary at the company; it was a point of bitter conflict between Ted and his father Samuel for years and the parents only married once the disapproving old man was dead and gone.

The Tootsietoy section of the Dowst catalogue soon boasted several vehicles, with models of the massive-selling Ford Model T proving particularly popular.

Not that cast toys were exclusively a US phenomenon. In 1893, London toymaker William Britain perfected the 'hollowcast' process for making toy soldiers. This involved pouring molten lead into a multi-part mould of a figure, swilling it around and then, as it was setting as a skin around the inside, pouring out the excess to leave a hollow metal soldier that could then be realistically painted. It made the raw material go much further and, of course, produced a finished product that was much lighter than a solid lead soldier. The Britains company became world leaders in this technique, before anyone made the link between lead toys (coated in lead-based paints) and the terrible poisoning hazard if they were sucked on by small children!

The realism of Britains' figures, which also included a range of farm animals, together with their excellent worldwide sales, must surely have been noticed at Meccano. Done at the appropriate scale, which would need to be somewhat smaller than Britains' 1:32, a similar range would make an excellent extension of the railway scenery paraphernalia that had grown up around Hornby Trains.

Although it's unlikely Frank Hornby had much hands-on involvement with their development – in 1931 he was 68, ailing from diabetes and busy with his political commitments – Meccano added its first Modelled Miniatures to its Hornby Accessories line-up. Its arrival was speeded by Hornby's decision, in 1932, to take his company public, which raised £300,000 in new working capital.

Modelled Miniatures were slush-cast from lead, a slightly simpler version of hollow-casting, and the first few issues included such layout essentials as station staff, farmyard animals, a shepherd set, passengers, and a cute lineside novelty advertising hoarding consisting of two workmen carrying a Hall's Distemper sign on their shoulders! They were of an extremely high quality and beautifully hand-painted. The figures were also available singly and the range was quickly augmented by a series of very small and simple cast trains intended for toddlers who were too young to handle building a layout and running the trains on tracks that needed to be assembled. One can fervently hope that the material used for these didn't cause too much brain damage ...

Modelled Miniatures was a strong-selling line for Hornby, drawing owners of train sets back into toyshops to add to their growing miniature rail networks. In December 1933, announced in the company's high-circulation *Meccano Magazine*, came the Modelled Miniatures set: the 22 Series of Motor Vehicles.

Just like all the other MM issues, they were cast in a lead-rich metal alloy and at the approximate 1:48 scale O gauge that made them natural partners to Hornby locos and rolling stock. The cost for the whole set was 4*s*, or they were offered separately for between 6*d* and 1*s* a pop.

No. 22a was a two-seater sports car, a generic representation but not unlike a Standard Avon roadster, while 22b was a Sports Coupé redolent of the rakish SS 1 – ancestor of the first Jaguar. Then 22c and 22d were respectively a pickup truck and a delivery van sharing the same chassis, while 22e was a farm tractor similar to a Fordson. Finally came a military Tank as 22f, dripping with fine cast detail replicating rivets and louvres, and boasting rubber tracks on six concealed wheels. Talking of wheels, these were lead castings like all the other small parts, apart from the radiator shell on the two cars, van and truck, which was a tiny tinplate attachment.

A Thirst for the First

A major Dinky Toy collection would have to include Modelled Miniatures as a bedrock, but deep pockets are essential these days. On 22 November 2011, auctioneer Vectis sold the exceptionally rare Sports Coupé and Sports Car from the original 1933–34 22 Series for £3,800 and £2,100 respectively, while in 2014 a complete line-up of 22 Series Motor Vehicles found a new home at £7,200.

Anyone going Christmas shopping for children in 1933 who set eyes on these little vehicles was no doubt impressed and beguiled by their fine detail and bright paint colours. From January to March 1934, they were still a talking point in the toy trade, although toyshops and department stores would have regarded them as simply another part of the Hornby offering for which display space would have to be found. That, however, was all about to change.

2

Runaway Success

Frank Hornby had the idea for a magazine devoted to the self-contained worlds his products created in 1916, when *Meccano Magazine* was first published. It was the chirpy organ through which, among the articles about how to build an Eiffel Tower from Meccano or real-life features on the locomotives aped by Hornby Trains, new products were previewed. So the April 1934 edition of the monthly was highly significant in introducing Meccano's 'Dinky Toys' to the world.

The recently released Modelled Miniatures series, briefly renamed Meccano Miniatures, now formed the first Dinky Toys series of road vehicles, but an idea of the ambitions for this new venture was provided by the announcement of the 23 Series of Racing Cars, the 24 Series of Motor Cars, the 25 Series of Commercial Motor Vehicles, No. 26 a single Rail Autocar, a Tram at No. 27, and the 28 Series of Delivery Vans.

The new products were outlined in brief and then rolled out throughout the year. In June, the No. 50 Series of Dinky Toys Ships and the No. 60 Series of Aeroplanes were announced. By the end of Dinky Toys' first nine months in Britain's toyshops, the catalogue positively groaned with some 150 different items.

However, what is especially impressive about this workload is that it was all achieved just over one year after Modelled Miniatures first appeared. And in that time Meccano put itself through a sudden and fascinating tangential change in where it was heading with die-cast toy cars. Why the hesitant launch of Dinky Toys and then the sudden blitz of new products? For the answer, we need to return to Theodore Dowst.

Dowst's Tootsietoys had been selling well all over the USA – so well, indeed, that the manufacturer of life-sized cars Graham-Paige teamed up with Tootsietoy with a plan to manufacture toy versions of its automobiles that could be a valuable promotional tool in the hands – literally – of sons of potential purchasers. It wasn't quite the first time that children-targeted persuasion had been harnessed by the motor industry; in the 1920s, Citroën had its own clockwork tinplate cars made on its behalf by CIJ. However, the Graham-Paige deal, we must presume, stipulated that there should be a range of models to represent all the different body styles offered on the real things. And it was probably for this reason that Dowst designed and patented a new die-cast concept.

This consisted of two castings. There was a standardised one for the chassis element that would include the base and the four mudguards and then individual ones for the various upper body options including limousines, sedans, coupés and delivery vans.

The bodies were designed to fit into the chassis, where the axles were then inserted from the side through matching holes in both parts to hold them together. These axles were essentially like nails, with the flattened head at one end holding the wheel in its position, and the pointed end machine-flattened to secure the wheel on the other side, and also to hold the whole vehicle together.

It was an ingenious and versatile piece of design work, and the two major components helped to spread the pressure on the toys when they were played with enthusiastically, meaning the castings were much less prone to breakage. And it's important to remember too that their realism was hugely boosted by tiny white rubber tyres on cast metal wheel hubs.

Tootsietoy's 'Grahams' were handsome and robust playthings and they were launched in 1933 in shops and as showroom giveaways. However, they didn't do too much for the Graham-Paige marque itself, which stopped making cars after 1945. Meanwhile, Tootsietoy extended the concept to a range of LaSalle models (LaSalle was a division of Cadillac), while also continuing to make cheaper die-cast cars as single castings, including Fords and Mack trucks. The toy industry was notorious for copycat products and Dowst would have made a stout legal defence of its patents. Its rival the Parker White Metal company, for instance, was tempted to mimic Tootsietoy's Grahams for its Erie-branded toy cars, but soon abandoned the idea, while Manoil swiftly modified its die-cast cars after its first batch of obvious Tootsietoy doppelgangers.

Over in Liverpool, though, the directors of Meccano must have seen the Graham range soon after it appeared and these Tootsietoys clearly caused panic at Binns Road as they represented a serious rival for Modelled Miniatures – especially because they were to O-gauge size like Hornby Trains. They were also of manifestly better, stronger design, not just in their construction method but because they were made from a relatively new die-casting alloy called zamak.

The name is an acronym of the German words for zinc, aluminium, magnesium and copper (kupfer), and it is overwhelmingly zinc with 3–4 per cent aluminium, 1–2 per cent copper and a tiny pinch of magnesium.

Meccano did nothing less than copy wholesale the Graham design for its 24 Series cars and much of the same concept for the 25 Series lorries. This was even down to the rubber tyres and separate radiator grilles. A version of the zamak compound was used but under the British name of mazak. Meccano must have gambled that Liverpool was beyond the reach of Dowst's lawyers in the USA, but it was still an audacious move and was no doubt done to pre-empt any licensed manufacture of Tootsietoys in Britain; to get in first. Their gamble paid off: no patent infringement suit was ever launched against Meccano and the Liverpool company even had the temerity to include a Town Sedan and an Ambulance in its 24 Series, blatantly copying Tootsietoy.

Dinky Toys were sold at between 6*d* and 1*s* apiece, and the motor vehicle series was also sold as boxed sets for between 3*s* and 6*s* 6*d*. What was so captivating about them was their detail – nothing like it had been seen up to that time in mass-produced toy cars of this handheld size. Of course, this was thanks to the intricate craft of the model maker who, with the die-casting process, was able to incorporate fine, almost hairline detail into the hardened steel of the moulds themselves. This would have begun with the hand-crafting of a wooden master model, with, for example, the door shut lines and bonnet louvres added in wire as the item made its slow process towards becoming the master for casting. All this intricate detail on the 24 Series Sportsman's Coupe, Town Sedan and Ambulance was picked out artfully.

Meccano was proud of its bright paint-colour palette for its new range and, although this could be chipped off when the toys were played with, the mazak casting underneath remained very sturdy. Or, rather, it should have done.

The quality of the molten metal in the alloy was crucial. There would always be traces of other metals present as impurities, but anything more than minute quantities could ruin the balance of the constituents and lead to a kind of metal fatigue called 'zinc pest'. Not that anyone could have known that at the time when production got under way. However, the company soon had to get a grip on the purity of the material to stop Dinky Toys' wheels from disintegrating or their window pillars cracking. It is thought that one cause of the fatigue that led to fine parts crumbling into granules may have been foil wrappers from cigarette packets, tossed casually into the molten metal by Meccano workers.

There were three main ways you could purchase Dinky Toys: from department stores, via mail order catalogues and – by far the most common – from dedicated toyshops that were existing stockists of Meccano and Hornby Trains. In the early 1930s, most significant towns had several toyshops that would be very well-known landmarks to local children. Proprietors would be visited regularly by representatives of manufacturers such as Meccano, who would extol the profit possibilities on offer from new and upcoming lines. No doubt the reps went into overdrive when demonstrating their bright little metal cars, lorries, ships and aircraft, and the shopkeepers must have been mightily impressed, because they were a runaway nationwide hit almost from the off.

One of the nation's most famous toy retailers, then as now, was London's Hamleys, and you can well imagine the impact that the big new Dinky Toys range made there in the run-up to Christmas in 1934. In 1931, the West End store had been acquired by Lines Brothers, bought from administration because Lines, which claimed to be the world's biggest toymaker, was its most prominent single creditor. Walter Lines, the human dynamo who drove this family business, had long been an admirer of Frank Hornby and must surely have seen the impact that Dinky Toys made.

Mass motoring was beginning to spread across Britain by the early 1930s, especially as the wide availability of affordable second-hand models made it possible for many British families to contemplate owning a car for the very first time. Fathers, mostly middle-class, white-collar and keen to swap a tram ticket or bicycle for four wheels, started to buy motoring magazines as they geared up to the big purchase. Then, when the noisy new machine arrived, their young sons and daughters would be beside themselves with excitement. In boys particularly, a passionate interest in cars would be ignited and the Dinky Toys' launch was timed perfectly to ride this wave of fascination, elbowing aside the Victorian lure of steam trains and First World War militaria.

Walter Lines saw all this and was among the first to take on Dinky Toys. Interestingly, he decided against investing in die-casting technology and instead adapted the tinplate expertise of his south London factory and workforce for the Minic line.

These were part of his company's vast Tri-ang range and, with their clockwork motors and larger 1:36-scale size, they were designed to sell for rather more than Meccano's Dinky Toys. Many were wonderful examples of the toymaker's art, shaped and crimped together with hand tools, and they copied Dinky Toys in offering a range of cars and lorries but with obvious similarities to real-life vehicles, such as Rolls-Royces, Vauxhalls and Fordsons. Some had opening panels secured with delicate wire hinges. However, the fact remained that they used old-fashioned technology and, with so many small pressed tin parts, they were susceptible to damage and malfunction almost as soon as they were out of the box. Bits would break off, the clockwork key would be lost; sons and fathers would be frustrated in about equal measure. Mass-produced Dinky Toys might have sustained chipped paintwork but they were built simply and, in general, pretty robustly because the detail was moulded in three dimensions instead of printed on two-dimensionally. Tri-ang's Minics did quite well in the toy market, certainly, but Meccano's Dinky Toys sold in truly phenomenal quantities. Industry organ *Toy Trader* lauded the product range for its realism, quality and vivid finish. And nor was their popularity restricted to the British Isles; a Dinky Toys range went into production at Meccano's factory in Bobigny, Paris, at the same time, or very shortly after, the molten metal started to flow in Liverpool. Some of the toy vehicles cast there were near-identical to their British counterparts, while others such as a fine rendering of the Peugeot 402 had specific appeal in the lucrative French market.

Once Meccano's new product line was in the shops and selling like hot cakes (they generated £50,000 in annual revenue in 1934, and £75,000 the year after), the company could take stock and fine-tune its offerings. It had already guessed, correctly, that schoolboys might like to collect all the items in a specific series if finances wouldn't allow the whole thing to be purchased in one go. Perhaps the craze for cigarette-card collecting among children had been noticed by the Binns Road bigwigs; certainly the twelve different and beautifully executed liveries of the 28 Series of Delivery Vans, ranging from Palethorpe's Sausages to Kodak Cameras, somehow impelled you to want to own the whole line-up.

If your father worked for, or was connected with, any of the firms arrayed in the vans' colourful transfers, then you'd naturally want to have that one. In 1935, Meccano boosted this with a range of Scammell Mechanical Horse articulated vehicles, with regalia derived from delivery and railway companies; one of these, presented as a refuse-collector truck, was the very first Dinky to feature an opening panel. Then came a brand-new range of 28 Series Vans with a host of new liveries including Atco lawnmowers and Hovis bread. Using other brand names to boost toy sales was a cunning ploy that was to last for the next fifty years and the proprietors can only have been delighted with the marketing messages that were implanted into the impressionable minds of future consumers.

So far, Dinky Toys had shied away from identifying their little metal models as replicas of actual vehicles. In 1935, though, the pretence was lifted, and the new 30 Series of Motor Cars included a Chrysler Airflow – one of the trendiest new cars around – a Vauxhall saloon, a Rolls-Royce sports saloon and a Daimler limousine. In fact, Dinky started collaborating actively with real-life manufacturers. For example, in October 1934, Daimler was asked for a set of photos of its Fifteen Sportsman car and a year later Meccano sent it a sample of the finished toy for approval. It was hardly a precision replica, but carmakers such as Daimler were quick to grasp the enormous publicity value these ninepenny, pocket-sized ambassadors could convey. Dinky Toys did the advertising for you, even if the potential customer was twenty years away from buying a real car.

That was just the start. In 1936, some 6,000sq. ft of extra manufacturing space was added to the Binns Road plant. As well as for making Dinky Toys, this was in preparation for the 1938 launch of Meccano's new Hornby Dublo, a whole new trains universe at 1:76 scale, following the same smaller-size lead of Germany's Bing. Dublo relied very much more heavily on die-cast parts than bent steel strips or tinplate, with an accompanying quantum leap in precise detail that was a direct benefit of the parallel Dinky operation. Also in 1936, Meccano launched its one and only attempt to appeal to little girls. The Dolly Varden doll's house – named after a character in the Charles Dickens novel *Barnaby Rudge* – was made from leather-covered boards that slotted together to form an expensive-looking suburban des res with a long drive along which Dinky Toys cars could be driven. It was supposed to be in scale with the cars although the range of furniture, all of which was die-cast and painted, was manifestly much bigger. It was one of the company's rare flops.

Meanwhile, over at the Dinky design department itself, staff had developed quickly a sharp sense for subjects that excited the public. The high-octane excitement of Grand Prix racing cars and heroes chasing the World Land Speed record provided loads of inspiration. In the 1930s, the Germans set the pace in Grand Prix racing and Dinky Toys was quick to include the Auto Union Type C and the Mercedes-Benz W25 in its range. One of the speed kings who regularly made Britain's national newspapers and radio news bulletins was Capt. George Eyston and Dinky modelled no fewer than four of his record-breaking machines – MG EX127 Magnette, a streamlined Hotchkiss, and the Thunderbolt and Speed of the Wind record cars. In 1939, they were joined by Lt A.T. 'Goldie' Gardner's MG EX135, a particularly fine and accurately finished representation of this 200mph machine.

Along the way, all manner of other vehicles were added to the extensive Dinky repertoire, including military vehicles, a caravan, motorcycles, a taxi and a Royal Mail van. The range, in just four years, had become enormous. Yet in mid-1939, as war clouds were gathering, the brand outdid itself when two series of cars were announced that set new standards of accuracy and attractiveness. These were the 38 Series of Sports Cars and the 39 Series of Saloon Cars, each one six-strong. They eclipsed everything Meccano's Dinky Toys had achieved up to then.

The company's model makers hit new heights in terms of accuracy and proportion and the detail extended to padding on the leather seats, separate headlights, steering wheel and acetate windscreens (although door mouldings were left off, weirdly) on the Sports Cars. Much of the same went for the Saloon Cars, all of which were American subject matter and included a Packard Super 8, Studebaker Commander, Lincoln Zephyr, Oldsmobile Six, Buick Viceroy and Chrysler Royal.

No longer did cars share the same generic, and therefore unrealistic, chassis and mudguards. Each body casting was recognisable from any angle for the car it was representing. Although the standardised wheels and nail-like axles remained, these series also did away with the old Tootsietoy construction method; the axles rested in lugs cast into the underside of the body and were then secured in position when a tinplate baseplate was riveted over them. And there was more: each car now had its name stamped into its baseplate, so any owner could be absolutely sure which miniature car he was playing with.

It was at around this time that the intriguing figure of Henry Hudson Dobson enters the Dinky Toys world. He's an obscure fellow, although by no means sinister, and what little is now known about his life has been pieced together by lifelong Dinky collector, and amateur detective, Terry Hardgrave. But without him Meccano's die-cast toy cars and trucks could not, it seems, have succeeded in the USA.

Quite how Hudson Dobson originally became associated with Meccano Ltd in Liverpool is unclear, but in 1913 and aged 22 it seems the former clerk was dispatched by the company to New York to open an office and warehouse so Meccano could start to export to the USA. It appears he was an effective organiser at Meccano Company of America Inc., and by 1922 was instrumental in setting up a short-lived Meccano construction kits manufacturing outpost in Elizabeth, New Jersey. He evidently had spells away from the firm to do his national service and work for a pottery company, but by 1933 and now a US citizen, Hudson Dobson was back at Meccano Inc., and in 1937 he sailed to Liverpool with a proposal. The upshot appears to be that he secured sole US distribution rights for Meccano, Hornby Trains and Dinky Toys, and established the H. Hudson Dobson Company in New York to handle them. With his long experience and contacts, Meccano could ill afford to lose his services, although the precise nature of the partnership remains a mystery.

We will probably never know for sure but it's highly possible Hudson Dobson, on that momentous trip urged the Dinky Toys people to produce the 39 Series of the very latest American cars, to give him something really outstanding with which to tempt young buyers and their parents. The timing all fits, and so does the fact that the construction method, with a sealed-on baseplate securing the cars' axles in their die-cast carriers, had evolved right away from the Tootsietoy body-on-chassis structure, which meant a legal challenge over patents would be deftly averted. With the new arrangements in place, the US market began to take an ever-larger proportion of Dinky output, and from that point on – with Hudson Dobson in charge – a distinctly American flavour to parts of the Dinky catalogue would be an absolute constant for the next forty years.

There were still very few rivals even five years after Dinky arrived. Toy soldier experts Britains manufactured only a few lead-cast vehicles, while Tri-ang's Minic series included just two 1:43-scale clockwork vehicles, a Ford Popular car and van, that were comparable in price.

One important Dinky Toys factor came into play in the mid-1940s: the increasingly warm reception awaiting them in export markets. In the USA, the sophisticated new metal models found a hungry market, especially as Tootsietoy now concentrated on simple, basic toys for toddlers rather than Dinky's replicas that delighted the detail freak in older boys and their dads. In 1939, Dinky Toys produced in England and France accounted for 16 per cent of company sales at Meccano Ltd, generating £81,000, and there were well over 300 cars, trucks, ships and planes in the range. Dinky Toys had soared in popularity, moving away from being mere accessories for model railway layouts to become essential to every car-mad schoolboy's life.

Dinky's 38 Series Sports Cars

This much-admired line-up was announced in summer 1939 and captured in small scale some of the most desirable open cars of the day. The series as announced included the Frazer-Nash BMW 328, Alvis 4.3, Sunbeam-Talbot, SS100 Jaguar, Lagonda V12 and Triumph Dolomite. Yet, because of the declaration of war against Germany, only limited numbers of the first three actually reached shops for the important pre-Christmas season. The Lagonda wasn't issued until April 1946 and the SS100 only appeared six months later. As for the Triumph, it was never issued at all. However, this didn't stop it from becoming something of a unicorn among collectors, the source of rumours and frenzy for its mythical existence. Fifty years later the mystique was still strong enough for a Dolomite finally to be included in the short-lived Dinky Classic Collection revival. No prototypes have ever been found. Back in 1946, its place in the 38 Series was taken by the Armstrong Siddeley Hurricane, which, incidentally, was the very first new post-war car to be modelled.

3

Dinky's Post-War Resurgence

Meccano's large Liverpool factory switched from making playthings that provided endless pleasure and boosted youthful imagination to the mechanised production of the grim tools of war. There would always be a certain poignancy to that. But, of course, needs must, and the company's three decades of accumulated expertise in metalworking meant its place in the 'war effort' was both assured and vital.

By 1942, all Meccano, Hornby Trains and Dinky Toys manufacture had been halted. The workforce was reassigned to a wide variety of tasks, which included building airframe parts for Wellington bomber planes and making bomb-release mechanisms. The huge number of women workers accustomed to assembling Dinky Toys on long production lines switched to producing other small but essential items that included die-cast parts, such as screws and fuses for shells, and even hypodermic needles.

As the end of hostilities approached, the firm found itself in possession of excellent plant and machinery as a result of the Ministry of Supply's exacting demands. It seems the factory itself sustained little bomb damage and most of the Dinky Toys dies had been mothballed carefully, so production could get under way again with minimal difficulty. Whether it was the machinery cranking back into life in super-quick time, unsold stocks from the early 1940s or a mixture of both, there were limited supplies of some fifty different Dinky Toys in the shops for Christmas 1945. Fans would have to wait another four months for the first actual new issues, in April 1946, the first being a model of a military Jeep.

In the company pecking order, Dinky Toys now took second place behind Meccano itself, but ahead of Hornby Trains. Its salesmen, who tended to deal directly with retailers rather than using wholesalers, pressed home the benefits of affordable toys that could sell steadily all year round – not just in the frenzied run-up to Christmas – and that, thanks to continual promotion and encouragement in the *Meccano Magazine*, were likely to be keenly collected.

That was the good news on the home front, anyway, together with the fact that UK imports from Germany – traditionally the global masters of toy manufacture – would be banned until 1950. For export markets, Meccano must also have been cheered by the announcement, in November 1945, that no export licence was needed for toys and so the company could start to pursue overseas sales. The way was clear for Dinky Toys to take off again.

To do this, there needed to be a constant flow of new issues and in 1947 the company got cracking, with several revised versions of pre-war cars and lorries and a model of one of the very latest British family cars, the Riley RM saloon. The major news this year was the launch of Dinky Supertoys, a line-up of six lorries (plus an aircraft, the Short Brothers' Shetland flying boat) that were much larger and more impressive than anything so far from Binns Road.

The bigger size and the fact that each one came in its own individual box like a Hornby railway locomotive (rather than 'trade boxes', from which six pieces were sold loose by the toyshop) were notable. These lorries were a significant stride forward in die-cast detail, with the vehicles' grilles modelled accurately and both the four-wheeled Guy Vixen and eight-wheeled Foden Diesel chassis coming with a range of different rear bodies – you could have the Foden as a petrol tanker, wagon, flatbed or a chain-side lorry.

At 10s each, the Fodens especially were taking Dinky into a new and higher price bracket. A Supertoy like this was something for the young enthusiast to aspire to for a birthday or Christmas present, perhaps a reward for working hard at school; the smaller, less costly models were more attainable from pocket-money funds with a bit of saving discipline. Nonetheless, although Dinky Toys had established themselves quickly as the toybox *lingua franca* of the era – everyday items for the likes of William Brown in Richmal Crompton's *Just William* books – they were generally beyond the means of less well-off children and their families.

Throughout 1948 and 1949, the reissues of pre-war Dinky Toys became fewer and the number of brand new releases swelled. Lorries were very popular, with smaller-scale issues joining the Supertoys, while a new Bedford OB-type cab was adapted cleverly into a tipper, an articulated wagon and a refuse truck with tinplate shutters that could be opened for added realism. In June 1948 a new range of farmyard models was cranked into life with a Massey-Harris tractor, while the Supertoys line-up was boosted

Dinky's Fantastic Fodens: Solid Investments

Eight-wheeled, four-axle Foden lorries, as modelled by Dinky, were among the first post-war issues to romp ahead in value among serious collectors – a measure both of their pleasing appearance and evocative 1940s/1950s aura. Vectis, Britain's leading specialist toy auctioneer, found they were still confounding their estimates in 2015. A yellow and green Foden Flat Truck, with a second type cab, sold for £1,320 in its box against an expectation of £6–700, while a corresponding Flat Truck with Tailboard fetched £1,800 against a similar estimate. A Flat Truck with Chains, first type cab, sold for £1,920, nearly four times its £4–500 forecast.

with construction vehicles like a heavy tractor and a dumper truck.

Cars were not overlooked and 1948 saw the arrival of a Triumph 1800 and a Standard Vanguard, which were pretty faithful copies of the new post-war cars you might have seen on Britain's streets. In 1949 they were joined by an Austin A40 Devon and an American Ford Fordor. There were more farm implements, Supertoy lorries and other Dinkys that 'did things', such as a Coventry-Climax Forklift Truck with a string-operated winding/lifting mechanism and a Motocart – a very strange agricultural three-wheeler, with tipping back end. And nor were our sisters completely overlooked: Dinky's larger-scale die-cast range of garden implements, including roller, lawnmower, wheelbarrow and hand truck, were perfect for the garden of any doll's house!

For the first four years of the 1950s, there were new Dinky Toys virtually every month. There was a seemingly limitless cavalcade of new cars, lorries, farm equipment and 'action' vehicles that sent young consumers' heads spinning every time they went past a toyshop window. Some of them were models of vehicles that were very much of the moment, such as the revolutionary Land Rover, the controversial Austin A90 Atlantic convertible and the striking Hudson Commodore. Just as newsworthy was Dinky's series of Grand Prix racing cars depicting British and European rivals in mostly authentic colours and liveries.

Yet many children preferred Dinkys that offered imagination-boosting features, such as the 1950 Commer breakdown lorry and, in particular, the 1953 issue of a Bedford Pullmore Car Transporter that could haul four Dinky cars, and could be had with a tinplate loading ramp to get two of them on to the top deck. This was one of Meccano's most consistently popular and profitable Dinky Supertoys, a stalwart of the catalogue for more than a decade. One attractive 1955 issue was especially local in its inspiration: a specific reproduction of the Land Rover Series I rescue vehicle that patrolled the Mersey Tunnel.

Dinky Toys were incredibly robust. Tri-ang had introduced plastic wheels on some of its cheaper Minic cars but Dinkys, rubber tyres apart, were rugged die-cast metal throughout. They weren't absolutely unbreakable, but they still provided pleasure after years of being played with. The quality of paint and transfers were excellent and once they got battered young owners loved repainting them, no matter how unfortunate the end result. Any early problems with mazak die-casting quality issues had been largely banished and tinplate parts and manual mechanisms for moving parts were all integrated beautifully. Minic's plastic-shelled, friction-powered cars seemed very shoddy by comparison, and while there were other die-cast Dinky rivals from Timpo and Crescent, their lines were generic and their detail crude.

Some dips in output were forced on the company because of outside events. One of the most serious was the Korean War from 1950, the global ramifications of which caused some serious raw materials shortages. Meccano coped with this by concentrating on sales of its smaller Dinky Toys models in the retail market; for a time it suspended its Supertoys sub-brand for fear of not being able to meet demand. To give added appeal to its smaller cars and trucks, many started to be available in individual cardboard boxes from 1952.

In 1953, though, the problems lifted and it was boom time. Demand was such that Meccano opened a second plant at Speke on Merseyside that year to ramp up production of the basic toy castings, which were then transported to Binns Road for painting and assembly. Ever larger quantities were being bought at home and also shipped to eager export markets in the USA and Commonwealth countries, none of which had indigenous rivals whose products had anything like the vibrancy and quality of Dinky Toys. Indeed, the brand had had much of the global market for die-cast toy vehicles to itself for virtually twenty years. This prevalence surely couldn't last, and it didn't. We need to head about 230 miles south-east of Liverpool to find out why.

4

New Pocket-Sized Thinking

Rodney Smith and Leslie Smith weren't even distantly related, but they'd been very good pals ever since they met at the George Spicer School in Enfield, Middlesex, in the 1920s.

After the Second World War, in which they'd both served (Rodney in the Royal Engineers, Leslie in the Navy) they were still bosom buddies. Leslie would often get off the train at Palmers Green, north London, on his way home from work and knock on Rodney's front door. He'd be invited in for tea and the two would have a chinwag about their lots in life. It was 1946; Rodney was 28 and working locally for Die Casting Machine Tools Ltd (DCMT) while 27-year old Leslie had had a variety of jobs since leaving school at 14, and was now balancing working for his father's building firm with employment as a carpet export buyer. Their demobbed lives were plodding along okay, but they had a bit of cash and a vague idea that they wanted to start their own business.

The two wannabe entrepreneurs could have got into importing rugs from Belgium. Instead, their conversations turned increasingly to Rodney's growing experience in die-cast engineering and the lucrative market for cast widgets essential to British manufacturing. DCMT churned out mountains of parts for electrical products and toys and business was going crazy, so the two decided, after months of mulling it over, that they too should grab a piece of this action.

Rodney had £90 in war service gratuities and Leslie had £200. Scratching around, they managed to amass £500 between them and on 19 January 1947 they formed Lesney Products as an informal partnership. The name was an amalgam of their forenames, and they chose 'Products' because, at this stage, they had absolutely no idea what they would be making.

Rodney resigned from his job and spent the next three months looking for premises. What he found didn't seem very promising but, at £100 a year rent, it was very affordable. It was a dilapidated pub called The Rifleman on Union Row, in the scruffy hinterland between Edmonton and Tottenham in north London. An old garage or shop would have been more suitable but Rodney simply couldn't find one within budget.

In the cellar there was space to install the hand-operated die-casting machine Rodney had bought cheaply from his former employer. The ground floor would be for general assembly and sorting activities. Meanwhile, the first floor would provide a couple of offices and a storeroom, but only after Leslie's builder dad had strengthened the floor so the building wouldn't collapse when the storeroom was packed with zinc!

It was the chilly winter of early 1947. Leslie was still in salaried employment and Rodney was on his own in the old boozer one morning, waiting for the gas fitters, when another friend, John 'Jack' Odell, popped by. Like Rodney, he'd also worked for DCMT in its tool room and likewise decided he could beat it at its own game. Jack was quite the rebel, having been expelled from school at 13 and then drifting from job to job as variously a van driver, cinema projectionist and estate agent's clerk but eventually finding he had a propensity for engineering that he put to practical use in stripping old army vehicles for salvage. His cunning plan now was to run a rival business to DCMT in the evening, and he'd picked up six ex-army die-casting machines in Luton for a bargain £60. However, his ruse to operate them in his mother's garage had just been scuppered by the local council, which said planning laws wouldn't allow this kind of light industry in a residential area. Now, Odell had stumbled across the solution to his problem and his half-dozen machines were crammed into The Rifleman's basement, too. What was more, Odell had an order for a few thousand string cutters, a razorblade in a cast housing.

'We just decided that we would all go into it together,' Rodney remembered. 'There was nothing complicated about it.' Lesney Products became a three-way partnership. And a sound one it was, too: a die-casting expert, an engineer and a capable administrator-cum-salesman. Business got off to a healthy start with an order from the General Electric Company for cast components, but by October 1947 the machines were silent as work dried up. The partners discovered this was usual for tiny companies in the supply chain to industry giants, and panic set in. Then a local manufacturer of dartboards asked them to make parts for some toy guns and handcuffs and the trio realised these apparently frivolous products could carry them through slack periods.

Leslie, the finance and admin guy, liked the steady cashflow but Odell, with his love of precision, realised he could be both productive and creative. He'd been a whizz at repairing tanks during the war and was used to improvising with ingenious fixes and substitutions when he was up against it. The partners' research into the toy market was decidedly unscientific but they were well aware of the massive success of die-cast Dinky Toys and Odell had a feeling for vehicles. Working quickly and largely on his own, he designed a range of simple vehicles. There was a road roller, a cement mixer and a Caterpillar tractor with or without a bulldozer blade. They were finely detailed and well proportioned.

With no real plan for how to sell them, Lesney's die-casting machines were set to work making batches of the required components during downtime between contracts and the assembled models were painted in a variety of bright colours. Trays of the little vehicles were then hawked around local shopkeepers in Tottenham. Many managers leapt at the chance to sell them; reps from Meccano were very sniffy about who they would bestow a Dinky Toys agency upon, but Lesney would sell its wares to anyone. And the appeal was that these metal newcomers were designed to be sold at a third of the price of Dinky equivalents. The company had a minor hit on its hands. The partners received a significant boost to their fortunes when buyers at Woolworths (the lowly chain being disdained by Meccano) got wind of the new Lesney series and put in meaty orders so they could be sold nationally.

This sort of business put Lesney on a sounder footing. It became a proper, limited company in March 1949 and moved to much larger and more suitable premises for a toy factory in Shacklewell Lane, Dalston, east London. By the end of that year, new vehicles had joined its range, including a horse-drawn milk float and a soapbox racer, and it came up with a cute toy elephant with a tinplate body and die-cast legs that, via an internal clockwork motor, could walk the animal along.

The partners were opportunists, increasingly adventurous buccaneers in business, and they began pondering on how they could exploit the forthcoming Festival of Britain in 1951. They hit on the idea of making an elaborate model of the Royal State Coach. It was scheduled for use at the much-heralded event and so a handheld version could cash in on the anticipated patriotic fervour. Odell duly designed the intricate tools to cast the vehicle and its eight horses. His earlier vehicle designs already showed his uncanny eye for scaled-down accuracy and gift for design, with plenty of delicate details such as the bolts picked out on the Caterpillar tractor, but he excelled himself with the fineness of both the coach and the animals' contours. Events, though, threw a spanner in the works.

The outbreak of the Korean War put a heavy restriction on the use of zinc in toy-making and the venture was suddenly unviable. Odell is said to have stored the moulds under his desk. Along with a drought of other contracts, Lesney Products looked to be on the rocks once again. Rodney reviewed his options and in 1951 decided that he wanted out. Decades later, he recalled:

Actually, I could see a future in Lesney. It was doing really, really well. Les and Jack were getting very big with their ideas, but for me I liked the cut and thrust of the start-up. Also, I had more interesting things to do. I had a half-share in a boatyard and I was spending quite a lot of time doing yacht deliveries; that was my great interest. I was living on a farm at the time and there was the chance of buying a smallholding. I spread my wings a little.

Rodney sold his third share in the company to Leslie and Jack and the £8,000 he received helped bankroll his many interests. Bearing in mind Lesney's phenomenal success in the years ahead, he must surely have regretted his decision. In fact, as early as 1954, his agricultural dreams had evaporated and he was back in the die-casting business on his own in one of his farm buildings. He was still friendly with his old partners and they gave him some of Lesney's redundant manufacturing plant. He then teamed up with London distributors Morris & Stone to produce the various Morestone and (from 1959) Budgie ranges of model vehicles. His business R. Smith (Diecasting) Ltd was a 50:50 joint venture with Sam and Harry Morris, but in 1958 Rodney elected to sell his half to the London brothers and emigrate to Australia. R. Smith (Diecasting) Ltd was wound up in 1969. He retired in 1982 and died aged 95 in 2013.

Back at Lesney, meanwhile, the downturn proved short-lived. The motor industry was unaffected by the zinc ban, so Leslie Smith went out and won orders for die-cast car parts that soon saw the company's Enfield satellite plant humming as it produced more than 1 million a week. For its toy-making activities, in 1951 Lesney teamed up with a long-established toy-trade distributor called J. Kohnstam Ltd, which took on responsibility for marketing Lesney's output. Then the Queen's Coronation in 1953 presented the ideal opportunity to dust off the dies for the State Coach and get the item into the shops in good time. It was a strong seller. In its own box bearing the Lesney name, the attractively finished model would sell 33,000 examples. But that was only the start.

Odell scaled the coach model down to a miniature version measuring just 4.5in long. This gleamingly plated coach was pulled by a train of eight, hand-painted horses; these were hollow-cast in lead using handheld moulds, and making them was contracted out to another casting firm, Benbros, in nearby Walthamstow; a company that then also decided to enter the die-cast toy market itself. The little Coronation Coach's success was spectacular and more than 1 million examples were sold through Kohnstam's supply lines at 2s 11d apiece. It was quite a brittle item and its tiny size meant it became a firm favourite as a cake decoration, those lead hooves alarmingly embedded in the icing!

It was intriguing that Lesney had seen such significant success with one of its smallest products, sold in sweet shops, novelty stores and other independent outlets. Smith and Odell were naturally keen to repeat the experience, but although they tossed ideas around constantly, inspiration came from an unlikely source. Odell's daughter Anne often took worms and spiders to school in a matchbox. Her father asked if she would stop doing this if he made her a toy to take into school that fitted inside her box. She agreed and in his spare time Odell crafted a tiny replica of the Lesney road roller, itself modelled on a real-life machine made by Aveling-Barford, fashioned entirely from brass and hand-painted in green and red. She took it in and, when he picked her up from school that day, he was besieged with pleas from her school friends for copies of it; her father was then on the cusp of becoming one of the greatest toymakers Britain had ever known.

He sensed there was another winner here for Lesney, and his partner Leslie Smith needed little persuasion. Odell proposed a series of tiny die-cast models of road vehicles packaged in boxes with the same proportions as matchboxes – specifically one he had to hand from the Czechoslovakian Norvic Match Co. Wholesaler Richard Kohnstam could instantly see the appeal to the legions of corner shops he supplied and that the range would be a natural fit for tobacconists as novelties alongside packs of Woodbines and tins of Old Holborn.

The Matchbox Series was launched in late 1953 and sold under the Moko brand, derived from Moses Kohnstam, the sales company's founder. The first four were the Aveling-Barford roller, Muir Hill Site Dumper, Cement Mixer and Massey-Harris Tractor. In 1954, the fifth issue was a London Bus, swiftly followed by a Quarry Truck, Horse-drawn Milk Float, Caterpillar Tractor and a Dennis Fire Engine. Sales began slowly. Some shopkeepers and wholesalers ridiculed the Matchbox toys as little better than 'Christmas cracker trash', but the arrival of the bus gave the range a huge fillip. The demand for these tiny, pocket-sized toys suddenly rocketed.

The reasons weren't hard to fathom. They cost just 1s 6d each yet still came in individual boxes bearing an attractive yellow and blue livery. And they were sold in non-toyshop environments, such as tobacconists, where generous adults could buy them on impulse to reward and delight their children if they'd behaved themselves. Conversely, if children were present then pester power would kick in and a grown-up would probably relent because they didn't cost very much and would be likely to quell any whining. Newsagents and sweet shops were other favoured outlets, for obvious reasons bearing in mind the target market among very young children.

Another factor was the subject matter. Although many of the first issues were actually scaled-down models of previous Lesney vehicles, all the early issues in the Matchbox Series were working vehicles, quite likely the sort of things that working-class dads drove for a living or that kids were familiar with from the streets, building sites and farms that formed the backdrop to their everyday lives. They were comforting and came in bright colours, often with radiators neatly picked out in hand-painted gold and silver by home-based outworkers, usually housewives or pensioners earning pin money and living within an easy striking distance of Lesney's factory. Some of them collected and returned their work in prams. And the fact that the vehicles were hopelessly out of scale with one another so that they could all fit inside identically matchbox-sized packs mattered not one bit. People might have called them 'toy cars' but, in fact, there was no actual car in the range until the 1956 issues of No. 19, an MG TD, and No. 22, a Vauxhall E Series Cresta. Even then, few of the adults buying Matchbox toys would have been car owners themselves.

The explosion in sales for Lesney's Moko Matchbox series might have caught lesser individuals by surprise. But Leslie Smith and Jack Odell were men of extraordinary, and complementary, capabilities. Leslie, who finally gave up his salaried job to concentrate on Lesney full-time when the business turned over £250,000 and employed 100 people, was an organisational genius. As demand soared, he formulated all the expansion plans, which included additional factory space, some of it at Stoke Newington, and the employment of hundreds of additional workers to manufacture, paint, assemble, finish and pack the Matchbox vehicles. Odell, once Smith was fully on board, ran everything on both the creative and the technical fronts. He was a manufacturer to his core but a man with a genuine love of the subject matter and an absolute zeal for accuracy.

Odell was also a gifted inventor. With demand for the Matchbox series exploding, the limitations of casting using hand-operated machinery were becoming clear, even though there were forty machines in continuous operation. Automated machines, while they did exist, were almost impossible to buy in, and there were fears for worker safety in equipment that pumped molten metal unstoppably at high pressure in a factory environment. Odell was driving back from a meeting one afternoon when, in his head, he schemed his own design with built-in safety features. He sped straight back to the factory that evening, ordered his assistants to tell their wives they wouldn't be home to supper, swept the floor and drew out the machine's workings in chalk on the concrete. The first vehicle to be cast in the finished machine was the Marshall MkVII horsebox, but the staff ran from the room just before Odell pressed its start button, terrified of being burned if it exploded. Odell's prototype automatic die-casting machine, however, ran like a dream. By 1966 he and his team had built 150 of them for the Hackney factory alone, fed by molten zinc from huge 6-ton furnaces, and some could make 7,000 sets of castings every day. These raw products turned out so strong and accurate that a very short time was usually needed to fettle them to remove flashing. What little over-matter there was, as for the very occasional reject, was smelted for reuse. The castings then travelled through the factory along Odell-designed automatic conveyors to be degreased, primed, spray-painted and then oven-baked. The fact that the creator of all this manufacturing apparatus was also the original designer of the models themselves marks Odell out as an exceptionally talented individual.

Indeed, it was Odell's reverence for the intricacies of very old vehicles that was behind Lesney's next launch. In 1956, the company unveiled its Models of Yesteryear range. Even the 'Yesteryear' term was an Odell invention, originally opposed by Smith and still detested by many, that soon slipped into the English language.

Odell found working on the smaller models, which he considered his bread-and-butter stuff, to be a chore. He longed to apply his skill at fine detailing to models of vehicles he admired and in a slightly larger scale. People in the toy trade had known for years that die-cast vehicles had huge appeal to adults as well as children and Yesteryears were tipped much more at the older enthusiast who would want to admire the skill that had gone into the model, rather than knock it about by racing it across the floor. In being aimed solely at collectors, the range was a world first for die-cast vehicles and at 2s 6d each they made very appealing gifts for difficult-to-buy-for chaps everywhere.

The initial Yesteryear range was an eclectic mix drawing directly on Odell's tastes for magnificent, British heavy engineering ... and with a definite aura of *Dad's Army* to it. Numbered 1–5, they were an Allchin Steam Engine, London Omnibus, Double-Decker Tram, Sentinel Steam Waggon and the 1929 Bentley 4.5-litre Le Mans winner. Once again, there was no consistency in scale; each model was created to fit inside a standardised box size. The next three in 1957 were an AEC Truck, a Leyland Truck and a Morris Cowley. The bus, tram and lorries all featured delightful period advertising. In 1957, Odell's design office produced the Yesteryear Fowler Showman's Engine, which to this day remains a die-cast masterpiece. It, like others in the series, sold in the hundreds of thousands and Models of Yesteryear played their part in awakening interest in veteran and vintage cars that led directly to the classic-car movement.

The range made its debut on the shared stand of Lesney and Kohnstam at the 1956 British Toy Fair. Only the very eagle-eyed would have noticed the Moko name was nowhere to be seen on the Yesteryear display and packaging. Its omission was a sign of the seething resentment that had grown between Smith and Odell and their distributor Richard Kohnstam. It went back to 1953 and the birth of Matchbox. Without the Lesney partners' knowledge, Kohnstam had registered the brand name himself. When they found out, Smith and Odell were livid and they never forgave this double-crossing even after he was forced to re-register the Matchbox trademark in all three of their names. Lesney distributed its Yesteryear range on its own in the UK and they went direct to US importer Fred Bronner with the new line even though he was originally an associate of Kohnstam, who was therefore cut out of the loop. It showed a steely side to Leslie Smith, in particular, which was as robust as the tiny metal car bodies cascading out of his firm's huge battalion of ever busier die-casting machines. Matchbox was on a roll: in 1961, it made more die-cast Yesteryear Rolls-Royce Silver Ghosts in one afternoon than Rolls had made cars in its whole fifty-seven years ...

5

Feature-Packed Corgi Toys

Among the galaxy of the nation's manufactured goods, die-cast toy vehicles made a plucky showing at the British Industries Fair at Earls Court in February 1956.

Matchbox, of course, was represented by its agent J. Kohnstam Ltd, of 393 City Road, London EC1, stating with typical ambiguity that they were 'producers of the Moko Lesney series'. Meanwhile, Meccano sought to ram home its market-leading status with an advert in the show catalogue that said its Dinky Toys were 'masterpieces in miniature. Solid metal; perfect finish'.

However, the real buzz was to be found on a stand carrying the banner of 'Mettoy Playthings'. Here was an all-new range of die-cast vehicles with an all-new brand name: Corgi Toys. And there was no doubt about their 'unique selling point'. The stand insignia screamed it and the industrialists and toy buyers could see it when they admired the gleaming samples of Corgi's Riley Pathfinder, Ford Consul and Bedford CA van. These really were 'The Ones With Windows'. If you examined a Dinky Toy car or van, it was obvious the interior was hollow and empty – the window apertures were gaping holes in the casting through which the unsightly muddle of fixings between the body, the baseplate and the wheels was clear to see. Corgi Toys, however, had clear plastic glazing and so they exuded an immediate leap in realism, enhanced by a superior level of casting authenticity. These toy cars looked uncannily like the real things in shrunken form. Naturally, Lesney's Smith and Odell were drawn to the Mettoy stand at the later Harrogate Toy Fair to examine them, and equally naturally they were hurriedly ushered away. 'It was the worst thing they could have done,' chuckled Jack Odell many years later, 'as it made us more determined to beat them and turn out cheaper and better stuff.'

Corgi was a completely new name in the toy market. Mettoy, by contrast, had been around for a long time, although even then few were aware of the unusual circumstances of its foundation.

The man behind Mettoy was Philipp Ullmann, who up until 1933 had spent twenty-one years building up the German toymaker Tipp & Co. in Nuremberg. It was a very successful business specialising in clockwork toys, of which vehicles were its mainstay, and many British children would have been familiar with its tinplate London buses and motorcycles because they were exported to the UK in large numbers.

After the Nazi party came to power in 1933, though, the outlook for Ullmann darkened overnight. He was Jewish and was convinced there would be trouble ahead for him from the ruthless and feared regime. An indication of just how much he felt at personal risk was that he fled the country without telling anybody, leaving even his secretary and chauffeur nonplussed, and headed for England.

Ullmann's reputation as a master toymaker was such that he found help easily in his newly adopted country to begin all over again. Bassett-Lowke, a Northampton-based maker of expensive model trains, welcomed him, and its engineering division, Winteringham Ltd, let him use its workshops. By 1936, he'd opened his own factory nearby in Stimpson Avenue, soon employing fifty people to produce a range of tinplate, clockwork models of cars and lorries, very much in the Tipp idiom. Mettoy was the company's name, a contraction of 'metal toys'. Its products were mostly exported to the USA and so didn't prove too troublesome to the domestic market leaders, Tri-ang's Minics and Meccano's Dinky Toys.

Joining Ullmann at the start of Mettoy was his cousin once-removed, 26-year-old Arthur Katz, who was born in Johannesburg and so possessed a British passport. Their families had lived in houses next door to one another back in Nuremberg and Katz fled Germany a few months after Ullmann, joining him in Northampton to become effectively the small company's first employee. They'd worked together at Tipp and now they made a tremendous success of Mettoy. At the outbreak of the Second World War, they presided over a 600-strong workforce at Harlestone Road, Northampton, and the company's metalworking expertise meant it became deeply involved in munitions manufacture – at first simple tinplate parts but soon shell and mortar carriers, Bren Gun magazines and grenade firing mechanisms among them. Mettoy concentrated on delivering for the Ministry of Supply, so much so that it was allocated another factory for manufacturing, a brand new 28,000sq. ft plant at Fforestfach, Swansea.

Of course, the manufacture of toys ceased entirely in 1941, but Mettoy hurriedly switched back to them from military hardware by 1946, now with its two factories in Northampton and Swansea. The success of Dinky Toys did not escape Ullmann and Katz. Their Mettoy catalogue, extensive as it was, lacked anything to rival Meccano's very strong sellers and as a short-term measure they bought in a small range of die-cast vehicles called Castoys designed and manufactured by the Birmingham Aluminium Casting Company. These were basic, generic models including a car, a fire engine, an articulated lorry and a van. They did include 'clockwork mechanisms' but the weight of the cast bodies meant these were pretty feeble performers. Nonetheless, they were a start.

Swansea seemed to present a new world of opportunity for Ullmann. It was near to the sources of metal raw materials, and there were plenty of both skilled and unskilled workers available locally for settled employment and good prospects, after the furore of the Second World War. With this in mind, in 1948 Ullmann took the bold step of building a huge new extension to his Fforestfach plant, erecting a 115,000sq. ft complex. It was opened by King George VI at a ceremony on 6 April 1949, and one of the first all-new ranges to emerge from the factory in 1950 were the Miniature Numbers, which included Mettoy's very first die-cast cars. These were pleasing models of the contemporary Standard Vanguard – in four different sizes, although the largest was a plastic edition – and Rolls-Royce Silver Wraith – two die-casts and two larger plastic iterations. They featured flywheel friction drive backwards and forwards, with the larger cars ingeniously changing direction if their bumpers hit an obstacle. Three years later, Mettoy added a plastic Ford Consul, too. Plastics became a major attraction for Mettoy, like most of its rivals. The company uncovered a major new market in polythene footballs sold under the Wembley brand and the mass manufacture of these soon kicked most of its other toy lines out of Northampton and across to Swansea where, in 1952, the factory was extended again to cover 200,000sq. ft and to house new departments to handle all the injection-moulded-plastics activity needed to make modern toys.

Those first Miniature Numbers had been designed by Eric Dixon, recruited from the Birmingham Aluminium Company, and it was Dixon who was in charge of installing all the die-casting and plastic injection moulding machinery that was leading the Swansea efforts. All the time, the company was making cautious steps into the die-cast toy vehicle sector, feeling its way and perfecting its manufacturing techniques. The impetus for a full-blown attack on the massively expanding market that was bringing rich profits for Meccano and now Lesney really began in 1953. Birmingham Aluminium lost another talented staff member to Mettoy as its head of development, Howard Fairbairn, defected to become a director at the request of Ullmann, his son Henry and Arthur Katz. Together, the quartet was about to target Meccano's Dinky Toys and the next three years would be spent planning their assault meticulously.

There would be one more crucial addition to the team: Marcel René van Cleemput. The 28-year-old Frenchman was taken on in Northampton as a designer on 1 January 1954. Van Cleemput had grown up in Yorkshire, graduated in engineering at Loughborough Technical College, returned to France to do his National Service, and had spent the last five years working for Express Lifts.

When the team analysed the Dinky offering, they found there was little consistency in scale and that the selection of subject was very random. They wanted their new models to have scale consistency yet also be of uniform sizes for packaging and perceived value reasons. So they decided the cars and vans would be in a span from 1:44 to 1:47 in scale, and lorries would be 1:56 scale – not exactly the 1:43 scale many assume. In addition, they decided the range had to consist of cars many people owned and were common sights on the roads of Britain. Not humdrum but still very familiar. Hence, the initial line-up consisted of the Ford Consul MkI, Austin A50 Cambridge, Morris Cowley, Vauxhall Velox E Series, Rover 90 P4, Hillman Husky and Riley Pathfinder. There were two sports cars, the Triumph TR2 and Austin-Healey 100-4. In addition, there were three variations on a Bedford CA van and two Commer lorries, a dropside truck and a refrigerated van in Wall's livery.

This was all a baptism of fire for new Mettoy recruit Marcel van Cleemput:

The first thing I was asked to do was design a little mould for part of a [toy] duck. Now, I'd never even thought of a mould before. I didn't know what a mould was like because I had been in lift design, press tool design and jig and tool design. But anyway, I soon caught on, and within six months I was chief designer. I did the drawings for the very first one and then I was involved in every single Corgi model that was ever produced until the company went bust in '83. Probably 95 per cent of the models emanated from myself. I was under Howard Fairbairn and he was a real hard driver, not liked by anybody because he was a bully. He was with me as soon as I came in in the morning and he would often keep me there until half-past seven at night! The pressure was there because we did two or three new models per month, and if each one had twenty parts then we had to produce twenty drawings, and then we had to produce the moulds for each model as well – each little block of steel in the mould had to be detailed, fully dimensioned; there's a lot of work.

The pressure was on to be better than Dinky Toys in every respect. The main innovation, of course, was the windows fitted to every vehicle in the range apart from the sports cars, formed with a one-piece clear plastic moulding fixed inside the roof of the body casting. In addition, they all had turned aluminium wheels that more or less resembled tiny chromium-plated hubcaps. Simple tinplate baseplates were fitted, riveted tightly into position.

Mettoy had for years favoured using flywheel-driven mechanisms – a precursor to friction drive – in its tinplate cars and lorries. This push-and-go feature gave them a life of their own without the need for a wind-up key as in a clockwork motor. It was decided to offer this in the new small die-cast range, so a parallel line of cars (and the Bedford van) was planned featuring 'motors' that allowed them to power themselves – up to 40ft over a smooth, flat surface, which was ample for most British living rooms. The same body casting would be used, but the mechanism would be carried on a die-cast base to make sure these issues, for which Dinky offered nothing even remotely similar, would be super sturdy.

They would all be stamped with the soon to be familiar legend Made in Gt Britain; it had to state this because, of course, the models would be made not in England but in Wales. And then there was the brand name for the range: Corgi Toys. It was suggested by Henry Ullmann, known internally as Mr Henry to avoid confusion with his dad, and was an allusion to the lively Welsh dog breed much favoured by the Queen. Corgi dogs were shown frequently in news pictures of the Royal Family. It was, perhaps, a typical case of immigrants truly appreciating the symbols of patriotism in their adopted country. The resonance was also not a million miles away from another certain trade name from Liverpool that sounded cute and suggested the product was small, desirable and tactile. The legacy of Tootsie Dowst lingered on ...

These toy cars were pretty much an instant success in the market when they reached shops on Monday, 9 July 1956. Corgi's bright packaging, eye-catching point-of-sale material for toyshops, a Corgi Model Club (with certificates, badges and newsletters, and administered by Mettoy advertising executive Bill Baxter) and, of course, the stunningly detailed and precisely proportioned models of the cars your family might own, had massive and immediate impact.

Much of the initial Corgi Toys success was down to the superb modelling skills of Marcel van Cleemput and his team. He was an intuitive sculptor. One of the clever techniques he perfected was to add a little extra width to the passenger compartments of the cars. He'd worked out that these would usually be viewed by children from above, as though they were gazing down at the cars from a high-rise block of flats, rather than at eye level as they would be seen on the street. Dinky's cars often seemed to have turret-like tops in which it looked as though there wouldn't be the space for the driver and passenger to sit side by side. Marcel's few extra millimetres made all the difference because super-accurate replicas simply looked too narrow to be realistic! The chief designer had an agreement from the start with his bosses that he could have one example of every model made and the fabulous collection he built up over the next twenty-eight years meant he could write the definitive guide to the history of the brand, *The Great Book of Corgi*, published in 1989. It's a uniquely brilliant book, packed with everything Corgi manufactured and described by the man behind every single design.

Van Cleemput's fastidious archiving meant he retained detailed records of sales figures and, thanks to this, we know that, in 1956, the firm sold about 65,000 examples of each standard saloon car and 45–48,000 of each mechanically propelled example. He also gave detailed figures of total sales for each model over their whole lives, and so from the first batch of issues we can see how they fared individually. For instance, the Hillman Husky would sell 352,000 examples over the next five years, while its mechanical version would sell 103,000 in three years (the M series were all dropped in 1959 because of their relative lack of success and higher prices). The No. 205 Riley Pathfinder sold 307,000 examples and the 205M version 110,000. And already it was clear that the more exciting subject matter sold best; the Austin-Healey, for example, shifted more than 500,000 copies, although it was on offer until 1965, much longer than most of the original line-up.

You've got to admire the way Mettoy executed its launch but also how it anticipated and delivered on the momentum it created. If you got the Corgi collecting bug in 1957 and were fortunate enough to be able to afford everything the company made, then Mettoy kept the new issues coming thick and fast, guaranteeing stockists that there would be something novel 'each and every month' from January.

Running through the year, first came a Karrier Bantam 2-Tonner Lorry, then the Standard Vanguard III and Jaguar 2.4-litre joined the saloon cars, and there was a new Commer platform lorry. The MGA was an additional sports car and then came a dropside trailer to match the Commer truck, a Bedford CA van in AA livery, a Vanwall Grand Prix racing car, the Land Rover and a Karrier Mobile Shop. That brings us to October, when the first Corgi Major model was issued, the Carrimore Car Transporter – a phenomenal example of model engineering for the time with a lowering upper deck and drop-down loading ramp (a massive improvement on Dinky's Pullmore, with its fixed upper deck and clumsy-looking separate tinplate access ramp). It could be loaded up with four of Corgi's cars. Rounding off the year were a Bedford CA Ambulance and the first foreign car to receive Corgi's exacting attention, a Citroën DS. There was also Corgi's first gift set – a natural mix of the Carrimore and four Corgi saloon cars. No doubt, if you got this for Christmas in 1957, you would be a very happy young man indeed.

The products themselves were bright and modern and the marketing was similarly up to the minute. Corgi issued its own catalogue for the first time in 1957 and the brand also harnessed the powerful new medium of commercial television. ITV had been on air for fewer than two years in Britain when, in March 1957, Corgi's TV campaign was announced. Mettoy took a spot on the first Sunday of each month to announce its new Corgi releases during Family Hour, which went out on the nationwide network. Plus, there were general Corgi commercials each Friday in the Midlands and, for a concerted month in April 1957, every Friday and Sunday across the country.

The company also got right to its target market of car-mad schoolboys by advertising in the *Eagle*, *TV Comic* and *Boy's Own Paper*. Meanwhile, at the coalface – the model counter at your local toyshop

– the point-of-sale material and the packaging couldn't fail to catch your eye, with a vivid blue-and-yellow theme, while the Corgi boxes themselves were resplendent with colour illustrations.

The final sales figures for 1957 – Corgi's first full year on sale – were mightily impressive, with 2,727,000 models sold, 911,000 of them going for export. The brand's attractive model of the Jaguar 2.4-litre topped the sales chart with 207,000 standard and mechanical examples sold. However, it was the Carrimore Car Transporter that really delighted the Ullmanns and Arthur Katz. At a retail price of 18s 6d, it pulled in six times as much revenue as a Corgi car and that spelt juicy profits for both manufacturer and stockist.

Corgi's new release blitz continued almost unabated in 1958. The new additions to the burgeoning range included a Mercedes-Benz 300SL roadster and Lotus XI, along with several RAF military vehicles including a Thunderbird Guided Missile and a Bristol Bloodhound Guided Missile, trailer and launch pads, and RAF Land Rovers to pull them around. On a rather gentler, more bucolic theme, some 183,000 examples of the Rice's Pony Trailer, with or without a Land Rover to pull it, found appreciative new homes with a few girls as well as the core boys constituent.

Corgi packaged some wonderful gift sets and, if intact, these have become highly valuable, with a Rocket Age Models Gift Set No. 6 auctioned for £2,040 by Vectis in April 2015. (At the same sale, a stunning Shell/BP Garage Gift Set No. 6, crammed with lovely vehicles, accessories and paperwork, went for £3,000.)

End-of-year sales totals massively increased to 3.2 million units sold, of which 1,040,000 were shipped abroad; in symbiosis, Corgi's first American car, the Studebaker Golden Hawk, both with and without mechanical motor, was a particularly hot seller with 199,000 examples sold.

From the very start of its Corgi venture Mettoy had secured a strong foothold in the potentially lucrative US market, thanks to the skill and panache of one Werner Fleischmann. Although hailing from Zurich, Switzerland, Fleischmann had emigrated to the States soon after the end of the Second World War, and for many years worked in the various enterprises of an acclaimed electrical engineer called Hazard Reeves. Among the sixty businesses Reeves founded were pioneering ventures in stereo sound for Hollywood's Cinerama movies, the manufacture of high-tension communication cables for laying on the Atlantic Ocean bed, and even the first prototypes that led to the creation of the fax machine.

In a 2021 interview, Werner's son Anthony recalled:

My father and Hazard Reeves did a variety of things together, and he worked with another man there called Homer Clapper who almost became my godfather. He was a pilot himself and he taught both my father and myself how to fly, which led eventually to me becoming a professional pilot for Swissair. My father and Homer loved automobiles as well as aircraft. My father had an SS 100 Jaguar, then later an Austin-Healey and an MG, and the two of them would actually race about town right here in New Jersey in those carefree times.

In the early 1950s, the restless Reeves decided to break up parts of his business empire, and his importing and distribution division Reeves International, founded in

1946, was acquired by Fleischmann after he decided to strike out on his own. With fluency in several languages, Werner devised a scheme to import, warehouse, distribute and sell high-quality European-made toys for the booming North American market.

'In those days, as I understand it, the toy business was really done once a year at Christmas in the department stores Macy's and Bloomingdale's, with packing cases delivered direct,' said Anthony Fleischmann. 'My father wanted to change this one-time-buy and facilitate the selling of toys all year round, as American wealth was growing. That then, frankly, allowed independent toy stores to come into existence.'

Werner made numerous trips to Europe and promoted his thinking there. In Germany he picked up exclusive North American distribution rights to top-notch model railways range Märklin and world-renowned teddy bears brand Steiff. In Britain he forged a close and long-lasting friendship with Dennis Britain, which means Britains' military and farm models were added to his roster, and it's through Britain that he was, most likely, alerted to the new Corgi Toys range, and in 1956 Fleischmann's Reeves International became the sole US importer for them too.

'These relationships went on for decades and I'm not aware there were any real official written agreements,' said Anthony Fleischmann. 'I've never seen any. They were handshake deals in the old-fashioned way. There was an affinity with the Ullmann and Katz families – they were all Europeans, like my father, operating outside their home countries. They understood each other.'

This brings us to the eve of 1959 – the threshold of Corgi's most dynamic decade – and that's a good point to consider what went into producing a Corgi Toy.

It took up to a year to take a model from proposal to manufacture. All the ideas went through Marcel van Cleemput as the range expanded. Once he'd deemed the model was a technical feasibility for die-casting, he'd contact the manufacturer for permission and access to original technical drawings of the subject, which were almost always eagerly provided, such was the new-found power of Corgi to generate interest in the real vehicles themselves. 'They'd always say yes. It was great publicity for them,' he recalled later for the Museum of Childhood's British Toy-Making Project. That was followed by four or five hours in and around the actual car or lorry, working with an assistant to measure, sketch and take up to seventy photos.

All of this went towards producing the 'general arrangement' drawings from which skilled model-makers would craft 'approval' models in wood or metal for Mettoy's executives in Northampton. At that stage, three out of four proposals were rejected. The drawings were amended and then carefully checked against the original vehicle once again.

Only when these plans were absolutely finalised would they be used as an accurate reference for making the dies. This process was extremely laborious, with lifetime-experienced craftsmen working away at the same piece of steel for several weeks and working at tolerances of as small as a thousandth of an inch as they produced the 'male' and 'female' interlocking parts of the tooling.

According to a 1959 report in motoring magazine *The Motor*, the finished tool would be tested by being used to make a pre-production batch of models to assess casting quality. 'Last-minute modifications on the actual car may have to be incorporated,' it recorded, 'and then the steel dies are hardened so they will last the life of the production run.' Meanwhile, paint finishes would be chosen and tested, sometimes using the colour choices of the original manufacturer of the real thing, or else an internal suggestion based on what would stimulate children.

That, though, was just the design process. The manufacturing was something else entirely. The internal working of a die-casting machine is naturally fascinating to engineers, but what it boils down to for the layman is being a mini foundry. The zinc-heavy mazak ingots were fed in at one end and the castings, several hundred per hour, came out the other, with the various die-cast parts formed on a tree-like metal frame called a sprue.

Each of these would then be removed from the sprues (which were melted down for reuse) by hand at break points carefully designed in. Wooden boxes would be piled high with one type of component, such as the body casting or the baseplates, which would then be taken to the factory's 'barrelling shop'. Here, great quantities of castings were placed inside gigantic, rubber-lined barrels – like tumble dryers in a laundrette – where, in a bath of fluid detergent and small synthetic pebbles called 'chips', they would be rotated gently. The smooth-edged chips, whose size was specially selected, would get inside every crevice and through every aperture of the casting to knock off flashing and other rough edges and unwanted casting remnants. With the soapy water and the drum's rubber lining, this would produce a smooth, clean finish in a process lasting about twenty minutes. Once they'd been through this the castings were inspected visually and any stubborn duds rejected as unsuitable to go forward for assembly.

A short documentary film, lacking sound or commentary, survives in the ITV Wales Cymru archives, showing the Corgi Toys manufacturing process from start to finish in its four minutes. It was filmed in 1960 and it's hard to tell whether the Swansea factory actually was as dark and grim as it's depicted, as the black-and-white footage is pretty grainy. *The Motor* said the assembly shop was 'light and airy'; I'm not so sure. What is immediately striking is the rigid separation of male and female roles. The blokes do all the heavy grafting, including the perilous handling and pouring of molten zinc, loading the tumbling barrels and hauling heavily laden trolleys around the factory floor, while almost all the painting processes and assembly tasks are undertaken by women. These ladies, apparently wearing their own clothes without overalls and sporting generally neatly coiffed hair without snoods or hats, are shown toiling away at tasks that can only be described as boring and repetitive – a veritable female army building toy cars for small boys who would have absolutely no idea that they were made for them by women like their own older sisters, mothers and, indeed, grandmas. You imagine – and certainly hope – that there was great camaraderie among them, and in 1960 the Fforestfach workforce was getting on for 1,000-strong in a plant that was now 240,000sq ft in area. It appears that the place can't have been anything other than full of metallic racket all day long.

You watch the washed and de-greased car bodies – in this case Corgi's Fiat 1800 saloon – being loaded on to wire grids for spray-painting by one woman, and then another feeding them into an oven where their enamel was baked to perfection ... and all without the skin-defending benefit of protective gloves.

The components are processed by more highly dextrous ladies, seated and standing, some of them operating stamping machines to punch out the Fiat's suspension leaves and plastic interior. The base unit is built up with axles and suspension at the various workstations and eventually, having been united with the upper body on a slow-moving conveyor belt, it's the job of one final section of the plant to fasten the two together and complete the process, with the Fiats whizzing down a final slipway towards the cardboard boxes that await them for packing and dispatch. In a fitting final shot, cartons of Corgi Fiats are wheeled away showing they're bound for Turin – birthplace of the real car.

According to *The Motor*, no more than six individual models would be in production simultaneously, but it was clear that Mettoy's Swansea factory was positively brimming with activity. And just look at the figures: in the year ITV shot its film, 6,248,000 Corgi Toys products were sold, a phenomenal 50 per cent-plus increase on 1959's 3.7 million total. This was all well before the all-time high points in the Corgi story, but 1960 did include one accolade that was a cherry on what was becoming a very large cake: Her Majesty The Queen bought three Corgi Land Rovers and Rice's Pony Trailers for her children, helping to make that, at 54,000 units sold, the brand's best-selling gift set of the year.

Frank Hornby, who added Dinky Toys to Meccano and Hornby Trains but didn't live to see their huge success. (David Upton)

Tri-ang's Minic toys were handcrafted from tinplate and featured clockwork motors but they were expensive and easily damaged.

Within a year of their launch, by 1935 the Dinky Toys line-up boasted dozens of different models to toast the new motoring age.

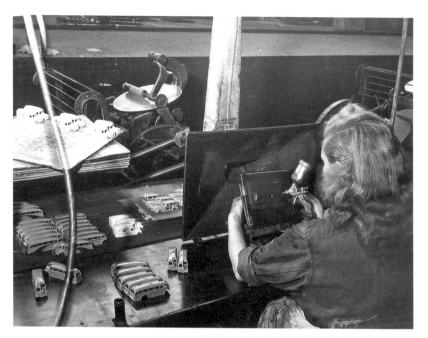

A lady working on Dinky Toys in the late 1940s, spraying stencilled detail on painted bodies of Daimler ambulances, Estate Cars and Single-Deck Buses.

Leslie Smith (left) and Jack Odell with a galaxy of the Matchbox models that made them east London's favourite millionaires. (Getty Images)

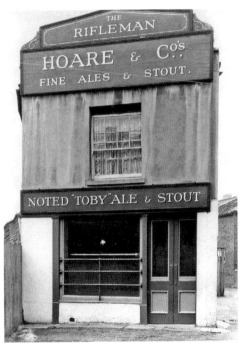

The Rifleman pub in Edmonton, north London, in whose dilapidated basement Lesney Products started its die-casting business. (National Brewery Centre, Burton-on-Trent)

Co-founder, co-owner and driving force behind Mettoy and, by extension, Corgi Toys, Arthur Katz CBE at Buckingham Palace in 1974. (Alamy)

The sprawling Mettoy complex in Swansea photographed in 1972, soon after a serious fire that destroyed a year's stock of Corgi Toys.

A batch of Corgi's second-generation James Bond Aston Martin DB5s on the tarmac behind the Swansea factory in 1974, a real Aston Martin DB6 in the background. (Alamy)

Another busy day at Mettoy's Fforestfach factory in the late 1960s as Car Transporter Gift Sets are carefully put together in their polystyrene packs.

Lesney's main factory at Lee Conservancy Road, Hackney, east London, a towering local landmark that ran red hot day and night. (Hackney Archives)

Lesney bought a fleet of second-hand buses to offer staff free travel to work, here at the company's Rochford, Essex, plant, the last to close down.

Free-spinning, shiny plastic wheels and a touch of hot-rod fantasy to the 1970s Matchbox Super Kings Tractor Transporter with two Superfast Mod Tractors as cargo.

A Dinky advert from a 1976 boys' comic to launch its excellent replica of the Triumph TR7, complete with 'working' deformable bumpers.

Pacey comic ad for Dinky's Eagles from the TV show Space 1999; *plans to transfer manufacture to Italy were scuppered when Airfix went bankrupt in 1981.*

6

A Universe of Miniature Dinky Transport Rolling out of Binns Road

In the last chapter, I walked you through the manufacturing process for Corgi Toys with the help of surviving contemporary reports. There were remarkably few differences in the way Meccano made its Dinky Toys, although the company did take the trouble to explain fully the process in a publicity handout in the early 1960s.

In talking about the finished dies, for example, Meccano gave a concise description of how the tool worked:

It is in two main sections which are mounted in an automatic die-casting machine for the actual casting. One section is bolted to the machine itself, and the other to a moving part called a platen, that can be moved horizontally to close the two half dies, while molten metal is forced into the space between them, and to separate them when this has been done. The metal solidifies almost immediately on entering this space.

The document also explained how the baseplates were held in place, using special spinning tools that flattened the end of a spigot inside the body casting once it had been inserted through a matching hole in the baseplate. To the uninformed, this simply resembled a separate rivet. It was the accuracy of this process that helped Dinky Toys to be such robust playthings. A child might just be able to bend an axle or detach a small die-cast component, but there was almost no way that the upper and lower parts of the car, van, lorry or bus could be separated to render them utterly broken.

This reputation for build quality, in addition to the vast range and attractive finish, had truly elevated Dinky Toys to a pedestal in the toyshop pecking order. Other toy vehicles had been judged against them and it was clear that cheaper rivals offered correspondingly worse value.

However, you do need to remember one thing about the toy market in the 1950s. Both the chief former foes of the Western allies had had significant toy industries; Germany was world leader in technology and quality, while Japan had specialised in cheap, efficient manufacturing. German toy imports were banned until 1950 and Japanese imports weren't allowed until as late as 1960. So companies such as Meccano had an unprecedented opportunity to exploit a market largely free of serious global rivals, most crucially in the USA. And in 1955, the Liverpool company was scooping £1 million in export sales annually.

The year before, Dinky Toys raced ahead of Hornby Trains and the constructible Meccano itself to become the company's best-selling range and would remain so for the next ten years straight. In 1956 the die-cast models accounted for 50 per cent of revenues.

To achieve this, Dinky Toys had doubled its sales between 1951 and 1957. That was the year sales hit an all-time high at 12 million models – with a million sent to the USA alone – and brought in £2.4 million to Meccano's coffers.

Dinky's transatlantic success, of course, was down to the methods of Meccano's British ex-pat importer Henry Hudson Dobson. His knowledge of the US market enabled him to persuade major department stores, other national and regional store chains, and independent toyshops to stock the British toy car range, supplied via his New Jersey headquarters. There was, inexplicably, no domestic manufacturer with anything even remotely similar. Every few years Hudson Dobson would travel to Liverpool to meet with his supplier – he also handled Hornby Trains and Meccano constructions sets – and no doubt he was treated like visiting royalty. The trips, by oceangoing liner, to Britain would last anything up to six weeks, and on one of them his wife Helen accompanied him, according to detailed research by historian Terry Hardgrave. It is my own speculation, of course, but perhaps the numerous Studebakers and Packards, as well as cars from marques including Cadillac, De Soto and Plymouth, that Dinky chose to reproduce in miniature arose from suggestions that Henry – a Jaguar driver himself, reportedly – tabled in the boardroom at Binns Road.

The overall gradual decline that followed this peak in Dinky's popularity saw British rivals bite ever deeper into the brand's market dominance. However, for the time being – and certainly as far as eager young consumers were concerned – the Dinky Toys bandwagon was rolling along like never before. From the mid/late 1950s to the end of the '60s, hundreds of new models were released, heralded each month in *Meccano Magazine*.

It would be fair to say, despite its huge and profitable output, those in charge of Dinky Toys' destiny had become rather complacent. The blame for that could reasonably be laid at the feet of Roland Hornby, Meccano chairman, who had experienced nothing but sales growth ever since he joined his father Frank in the business. Mettoy's launch of Corgi Toys did help to jolt the company out of its torpor, but even then Dinky seemed slow to react to the startling pace of innovation from Northampton and Swansea.

During 1956, Dinky added to its line-up of exciting modern sports cars, with a Triumph TR2 and Aston Martin DB3S joining its MG TF and Austin-Healey 100. The company's designers then turned their attention to alternative finishes, with the sports cars offered in both racing and road-going schemes. While they were at it with the paint samples, they also decided to give many of the existing saloon cars in the catalogue a makeover with two-tone paintjobs. A lot of these were lurid combinations, like the Austin Somerset in bright red and yellow or the Jaguar XK120 in cerise and turquoise, that would never be seen on the real thing, but it was a quick and cost-effective way to breathe new life into products that now had to do battle with 'The Ones With Windows' from Corgi.

The first fresh, competitive marketing move Dinky made was to start a Dinky Toys Club in January 1957 to foster loyalty, with its enamel lapel badge and membership scroll. It cost a shilling to join and the club played its part in helping sell some more of the new issues that year, which included several new small vans in various liveries, extra-large Supertoys lorries and a Mercedes-Benz W196 and Jaguar D-type to join the slightly more esoteric Connaught B-type and Bristol 450 Le Mans racing cars that appeared the previous year.

On a separate front, Meccano decided to have a pop at Lesney's Matchbox series with a new range of very small die-cast vehicles in December. The company hedged its bets by allying these to its model railway products under the Dublo Dinky banner, as they could all be used to heighten the reality of a OO-gauge railway layout. They broke a tiny bit of new ground in being the first Dinky Toys containing any kind of plastic parts, in this case their grey plastic wheels, but the Dublos were also-rans to the vibrant Matchbox line-up.

It wasn't until April 1958 that Dinky made a direct response to its Corgi aggressor. Its good-looking model of an Austin A105 was enhanced with clear plastic windows of its own – something that would feature on all Dinky's cars from hereon in ... apart from a Fiat 600 and Austin A30 in May and June, which were unglazed but had one-piece grey plastic wheels (dreadful – the wheels were undersized and looked especially crude). After this, Dinky Toys introduced regular new design features, which tended – it seemed – to trail their introduction by Corgi.

Hence, in December 1958, the Dinky De Soto Fireflite sported more authentic spun aluminium wheels (Corgi had them in 1956); in May 1960, the Standard Atlas Kenebrake minibus had a moulded plastic interior with steering wheel (Corgi: in July 1959); in October 1962, the Plymouth Fury had a removable plastic hardtop (Corgi: in March 1962); in February 1963, the Morris 1100 had an opening bonnet with a detailed die-cast engine inside (Corgi: in March 1960); in August 1962, the Jaguar Mk X had an opening boot (Corgi: in April 1961); and in August 1963, the Holden Special had tiny 'jewelled' headlights made from reflective cut-glass beads (Corgi: in April 1961).

However, it could be argued that these innovations – and the need to get in with them before the opposition – mattered more intensely to those in the industry than to the end user. All apart from the most fanatical collectors wouldn't have been keeping such a detailed score. And, it should be added, Dinky Toys' cars scored quite a few 'firsts' of their own.

In February 1959, the Dinky Toys Rolls-Royce Silver Wraith had sprung suspension (Corgi had this in October 1959); in August 1960, the Jaguar Mk 2 3.4-litre had 'directional control', a type of fingertip-operated steering (Corgi: in April 1961); in October 1962, the MGB had opening doors (Corgi: in November 1963); and in May 1963, the Ford Cortina Mk I had front seat-backs that tipped forward (Corgi: in November 1963). Sometimes, Dinky did something Corgi never tried, such as the seatbelt that could be unclipped from the lady driver of the Triumph Spitfire in September 1963, and front and rear passenger doors that could be opened on the Mercedes-Benz 600 Grosser limousine in November 1964. On top of that, the company was quick on the draw in 1962 with a Renault Dauphine Minicab, cashing in on a contemporary media furore surrounding the introduction of the rivals to London's black cabs.

Well before this eruption of engineering creativity, the popularity of Dinky Toys was experiencing a serious sag. Whether it was changes in public tastes or retailer attitudes, or intense pressure from Corgi Toys on quality and Matchbox on price – or a cocktail of all of this – Dinky's fortunes went into reverse. There was a huge shock in the USA. After nearly five decades' association with Meccano the man behind Dinky Toys'

North American success pulled the plug. Importer Henry Hudson Dobson made his final journey to Liverpool in January 1959, and shortly afterwards the partnership was terminated. The precise reasons are unclear but Hudson Dobson was then 68, and probably wanted to retire after winding up his business rather than see it go bankrupt (he subsequently lived until 1975). Presumably neither of his two sons, who had both worked in the family firm, wanted any further part in it either, and for the next few years US distribution of Dinky Toys was in the sporadic control of several regional companies as many stores and shops decided their precious retail space was better used for other, more desirable products.

One area that Dinky should receive recognition for was in its clandestine working relationships with the carmakers themselves to put a die-cast model in the shops at the same time as the real car was either being unveiled or reaching dealer showrooms. The first instance of this was for the launch of the Triumph Herald. Dinky's neat green-and-white model was in the shops just a few weeks after the wraps came off the full-sized car, which meant Dinky designers had been entrusted with confidential information long before Triumph disclosed anything about the Herald to the general public. Meccano's collaboration with the British Motor Corporation in 1962 was even closer. The toy manufacturer managed to have models of the MGB in shops on the very same day that the real car was launched in October 1962, and on top of that it was the first die-cast car made in Britain with opening doors (albeit ones that didn't fit particularly well). In November 1963, further hush-hush partnerships were revealed when toyshops received stocks of Dinky Toys' Triumph 2000 and Hillman Imp – both exciting new cars from British manufacturers – actually before the full-sized vehicles were on sale. What's more, both were packed with features that included an opening bonnet and boot, and working steering.

Of course, cars weren't the only story. The Dinky Toys range stretched out in all vehicular directions, although its military and agricultural ranges were mostly evergreen issues that soldiered on from their debuts in the 1950s. Dinky's roster of lorries, buses and emergency vehicles was added to constantly throughout the 1960s. They were big, colourful models with features such as tipping backs, opening doors, articulated trailers and the ability to interact with other Dinky models by pulling or carrying them. There was a tremendous amount of play value to be had from, for example, the Turntable Fire Escape of 1958, the 1959 Mighty Antar Low Loader with ship propeller cargo, the 1964 Brinks Armoured Car, and a range of vehicles themed around television outside broadcasting. Flashing, battery-powered roof beacons arrived on a Cadillac ambulance and an Airport Fire Tender in 1963, followed by flashing indicators on an impressive Vega Major Luxury Coach a year later; the boxy shapes of all three made accommodating the battery inside easy.

A comforting familiarity came with Dinky Toys that all young fathers could relate to in the early 1960s. They'd played with them when they were children and there was a natural continuity in that their sons were now keen on them, too. That ensured very widespread brand loyalty and a deep-seated belief in the quality of the range. And, of course, there was a continuously updated vista of new small wheels in the shops. Everything seemed as optimistic and go-ahead as it had done throughout the 1940s and '50s. Dinky was an absolute fixture of childhood, both at home and abroad. The reality, though, was that by 1963 the Meccano company was in a mess.

A variety of factors, including retooling costs and badly timed diversification (Meccano bought the plastic construction toy Bayko in 1959, a year before Lego was launched in the UK to much fanfare), saw losses soar from £10,000 in 1961 to £250,000 by January 1964. And that was despite slashing the workforce from 3,022 in 1960 to 1,507 in 1962. However, the problems were more deep-seated; Meccano was old-fashioned, inefficient and slow moving. Despite the prestige of its name and reputation, the *Financial Times* lambasted the company as 'the lame dog of the toy industry'. Historian Kenneth Brown outlined the situation succinctly in his book *The British Toy Business*: 'The long success of the construction set, the railway system and the Dinky cars had engendered a fatal lethargy in a firm which prided itself on high quality workmanship, irrespective of costs, and whose managers were largely recruited from within.'

And the lamentable thing about the situation was that the Hornby family was all too aware of it. In 1964 Roland Hornby paid a visit to Lesney Products' factory in east London and, after marvelling at the bustling efficiency, set about trying to persuade Leslie Smith and Jack Odell to buy Meccano and turn it round. But, of course, they had no need to saddle themselves with the problems of Binns Road, and showed him the door.

Lines Brothers, owners of the Tri-ang toy-making empire, sensed a bargain, and put in an audacious offer of £781,000. It must have been a measure of the distress Meccano was in that the shareholders, including the Hornby family who still owned 14 per cent of the business, accepted meekly. And this was despite the offer placing the company at less than half its then-current stock market value. On 14 February 1964, the deal was done and Roland Hornby along with his sister-in-law Una departed the boardroom with immediate effect.

The old guard had passed but ironically for Dinky Toys – the true bright spot in the Meccano gloom – it was probably the best thing that could have happened. Of course, the grandaddy of die-cast vehicles had not fallen to one of its upstart new rivals but to a new owner with a portfolio into which the brand could comfortably slip. In fact, Lines' Tri-ang division did have its own die-cast range, Spot On (see Chapter 9), but these highly detailed models never really caught on. Now with Dinky in its permanent grasp, Tri-ang decided to phase out Spot On.

Spot On cars were all to a rigorously constant scale of 1:42. It was a telling yardstick. By comparison, many of the Dinky cars issued in the late 1950s and early '60s were manifestly smaller than the 1:43 scale many assumed them to be. Now, and almost simultaneously with the Tri-ang takeover, new issues of Dinky cars and other vehicles looked and felt more substantial. The design ethos of range consistency and 'bigger-is-better' was a direct infusion of product-led vigour following the Tri-ang takeover. It probably coincided with the arrival at Binns Road of Graeme Lines, son of Tri-ang founder Walter Lines, as managing director, and no doubt there was also input from his cousin Richard Lines, a cheery innovator who was as energetic in his era as Frank Hornby had been in his. Product development was Richard's metier, exemplified by his Tri-ang OO-gauge railways system that undercut Hornby Dublo with its widespread adoption of plastic, and the Sindy doll range that mounted a spirited catfight with the all-conquering Barbie from the USA.

Tri-ang's Spot On series may never have troubled Dinky overmuch, but it had been started from scratch and used innovations to try and snare young buyers' fancy. The determined influence of the second-generation Lines dynasty promptly began with a 1:42-scale revamp across the board.

The scale of 1964 issues including the Rolls-Royce Silver Cloud III, Lincoln Continental and Mercedes-Benz 600 (which, at sixteen continuous years in the Dinky catalogue, had a longer lifespan than almost any other issue) was bigger; it was never stated by exactly how much – and lacked the stringent constancy of the Spot On discipline – but appeared to be between 1:42 and 1:38. They combined more opening features than ever, so to hold the models together die-cast baseplates were used instead of the flimsier-feeling tinplate that was common to Dinkys until then. The upscaling move continued with commercial vehicles, such as the Bedford TK Crash Truck and Coal Wagon; the AEC Articulated Lorry, Petrol Tanker and, later on, Hoynor Car Transporter, and the Mercedes-Benz Truck and Trailer. These all made earlier issues that lingered in the range as the 1960s rolled on, such as the Morris Mini Traveller, look very puny. And so in some cases, where a model existed in the older, smaller scale with fairly basic features, such as the Ford Corsair, Volkswagen Beetle and Rolls-Royce Phantom limousine, Cadillac Superior ambulance, Jaguar E-type and Ford Cortina, the Binns Road designers gradually created larger-scale replacements that were a vast improvement in accuracy, finish and cute detail.

Almost immediately there was an upsurge in sales. This was accounted for mostly by the US, where a change of distributor to Lines Brothers in 1964 (US firm AC Gilbert had held the concession only since 1963 and barely made any headway) stimulated stronger demand. In fact, Meccano couldn't produce Dinky Toys fast enough ... and were sometimes berated by the US salespeople at Lines Bros' New York office for that very inability. Back at home, the picture was less rosy. This was partly because of Meccano's old-fashioned notion of only selling Dinky Toys through approved outlets. A 1962 report by advertising agency J. Walter Thompson calculated that there were 6,500 places across the country where you could buy Dinky Toys but that rivals Corgi and Matchbox (sold in thousands of newsagents' and tobacconist shops, remember) could be had at 23,500 outlets.

Lines Brothers decided that quick and controversial action was needed to boost Dinky's market coverage – moves that would entirely bypass the old boys at Binns Road. New products were needed in a hurry and there was no time to waste on the usual, plodding development methods. The company felt that the main thing undermining Dinky's former dominance was the phenomenal impact that Lesney's Matchbox series had had on the market for young fanatics of model vehicles. Selling at just 55 cents each in the USA, the result had been seemingly ever surging demand as millions upon millions were lapped up. There were 75 to choose from, too, which made the buying and collecting of Matchbox cars, trucks and buses a low-outlay constant activity all year round – always something different to deliberate over before handing over that week's pocket money. But how could Dinky possibly hope to match that? All its models were big, heavy and detailed, and for all but the most spoilt brats had to be saved up for.

The answer was a range that had to have an edge on detail and yet still be price-competitive, and the solution was found in slashing manufacturing costs by sourcing them from ultra-low-wage Hong Kong. Frank and Roland Hornby would no doubt have been horrified at such a move, but their thinking was from another era entirely; it was imperative to reduce overheads and stand up to booming competitors. Moreover, the design would be tackled in double-quick time in 1965 in south-west London at Lines Brothers' Merton headquarters, where a line-up of 1:65-scale cars and an even smaller-scale range of road-building machinery was created. There was a 200-strong toolroom department there to oversee the work. The cars were an outwardly pleasing, authentic international array – from a Jaguar E-type and Mercedes-Benz 230SL to a Fiat 2300 station wagon and a Chevrolet Corvette Sting Ray; it seems likely the same engineers and designers who undertook Spot On R&D also worked on these.

Each featured at least one opening feature and sprung suspension, and the spun aluminium wheels with tiny, separate rubber tyres added a new level of sophistication at this size. There was novelty in the packaging, too, as each one came in a clear-sided plastic garage with an up-and-over door. Priced at between 50 and 59 cents in the crucial US market, the Mini-Dinky range seemed to offer a lot of bang for your half-a-buck in late 1966.

Quality, on the other hand, was shameful. Even in the shops, it was clear to see through the little plastic garage boxes that the opening doors and bonnets mostly didn't fit properly, and that the paint finish and thickness was very patchy. Thanks to poor quality standards in the die-casting process, they suffered from the sort of metal fatigue that hadn't been seen since the 1930s. The reception from stockists was decidedly chilly, but by that time Lines had already committed Dinky to its next Far East outsourcing venture. This time it was an all-new line-up of six slick cars that were the pride of the contemporary American suburbs, comprising a Buick Riviera, Chevrolet Impala, Chevrolet Corvair Monza, Ford Thunderbird, Rambler Classic and Dodge Polara convertible. It was no coincidence that the scale was 1:42 because this group had originally been planned (again, in Merton) as Tri-ang Spot On releases, but now the blueprints and prototypes were shipped off to Hong Kong too, with mass production contracted out to HKI, Loh Te-Sing's Hong Kong-based manufacturing company.

Once again, though, the quality was markedly second-rate in the fit of parts and the shoddy overall finish. One car, the Dodge, never even reached production and was replaced by an Oldsmobile 88. Lines Brothers Inc, just as for the Mini-Dinkys, found a lukewarm reception in the US; the 'Made In Hong Kong' statement on their flimsy boxes spoke volumes. Only two of the cars were sold in the UK, the Impala and the Riviera, and they were kept well away from the main range by being flogged cheaply and discreetly in chain stores like Woolworths. Like the Mini-Dinkys, they never made an appearance in any Dinky Toys annual catalogue ...

The plug was pulled in 1967 on any more Hong Kong product, which may have left some collectors tantalised at the multiple new issues listed on in-pack Mini-Dinky leaflets – Rover 2000, Aston Martin DB6 and Volvo P1800 among them – that would never actually go on sale. No doubt there was some schadenfreude relished in Liverpool at the demise of the cheapjack scheme from their London-based masters. But after that, Meccano itself designed a second line-up of

American cars at a similar size – attractive models including the Pontiac Parisienne, Chevrolet Corvette Stingray, Mercury Cougar and Cadillac Eldorado – and made them in-house to its customary high standards. The spectres of slow reactions and high costs would be back later to haunt them, though.

The 1968 new product offensive gained added shine with Dinky's adoption of electrically powered lights for its cars. Actually Tri-ang's Spot On was the trailblazer here. Its very first issue in 1959, a Ford Zodiac MkII, had the option of battery-powered headlights using an AA battery pressed into a slot through a gap in the baseplate casting. It was followed by a Spot On Rover 3-Litre with the same feature, and now it came to the Dinky line-up in January 1968 with a Mercedes-Benz 250SE whose rear brake lights glowed red when you pressed down on the rear suspension. It was soon joined by a BMW 2002 Tilux with battery-powered flashing indicator lights front and back, and then an NSU Ro80 with working head- and tail-lights, and luminous plastic seats. This Germanic trio were great fun just so long as the feeble internal connections stayed intact. The NSU in particular had superb proportions, while the Merc and BMW were a little top-heavy as the battery housing corrupted their lines somewhat.

A report in *Meccano Magazine* in August 1968 – written by the ever-ebullient Chris Jelley, the title's regular reporter along with colleague Mike Pedder on new Dinky issues – stated that each new model cost between £10,000 and £15,000 to tool up for. Ten years later, this had risen to £25,000. I know this because that was the figure I was given when I later wrote to Binns Road, imploring them to issue a model of the 1976 Aston Martin Lagonda; they politely declined my suggestion.

The mid-to-late 1960s was the period when pedantic young suckers for authenticity could find plenty to satisfy themselves. In 1964, for example, an agreement was reached with tyre manufacturer Dunlop for the company's lettering to grace the moulded sidewalls of the rubber tyres on Dinky's road vehicles. The Mercedes-Benz 600 limo was the first issue to sport them. In 1966–67, meanwhile, the Dinky design department gained yet another behind-the-scenes access opportunity to create a new model that would be ready for simultaneous unveiling with the real thing. They named it 'Dinky Model X' while in secret development, with a title guaranteed to arouse snooping. This time it was the Ford Escort two-door saloon, which, among its twenty separate components, featured opening doors, bonnet and boot lid. Few would have known that the real car and its approximately 1:42-scale reproduction would be manufactured just a few miles apart from one another in Merseyside, because Ford's bustling plant was in Halewood.

Although there was strife behind the scenes, the young consumers of Dinky Toys continued to lap up the new issues. The company had a massive hit on its hands when it ventured belatedly into TV character-based merchandising in partnership with children's TV legend Gerry Anderson (see Chapter 14). And its non-car releases tended to be full of interest, too. Highlights were the 1966 Ford Transit on which all the doors – including sliding ones on the side – opened (a Police Accident Unit version was even more fun); the 1966 Leyland eight-wheeled tipper whose cab tilted forward to reveal its power unit, as in real life; the enlarged 1967 Superior Cadillac Ambulance with a flashing roof beacon; the Jones Fleetmaster Cantilever Crane with its ingenious foldout mechanism; and the 1969 Merryweather Marquis Fire Tender with its own wind-out hose. All of them possessed a chunky, quality feel and would give hours of fun on the living-room floor. Little boys would spend whole afternoons gazing at the illustrations of them in the annual Dinky Toys catalogue and deliberating what to ask their parents for as their next birthday and Christmas presents. In 1968, for example, a new military Jeep pulling a Howitzer gun and a Volkswagen *Kübelwagen* extended the Dinky reach into other toymakers' territory because they were to a new 1:32 scale fully compatible with toy soldiers produced by the likes of Britains, Timpo and Crescent. These were, in fact, another legacy from Tri-ang, which had created the models for a new series of its own to be called Battle Lines but which, at the last minute, was folded into the Dinky range.

In the late 1960s many Dinky Toy cars started to be sold in impressive Perspex cases, mounted on plastic display plinths using holes cast into their bases. They looked fantastic – like precious museum pieces – and this meant that the increasing number of small 'chrome' plastic parts such as wing mirrors and aerials couldn't be broken off in transit (although they inevitably were when played with for the first time). I vividly recall seeing these cases in 1971 when I was 6, at a birthday party for a rather indulged only child. In his playroom just about every Dinky car then on sale was displayed in its case on the windowsill and a shelf – all apparently brand new, pristine and unplayed-with. I was absolutely green with envy at this galaxy of gleaming metal, but our host forbade us in strident terms from so much as touching them! You couldn't help but feel Dinky, at least from the end user's viewpoint, was at its very finest.

7

Matchbox 1–75 Series and the Industrial Juggernaut of Hackney's Star Enterprise

It was a Ford Thunderbird that sealed it, a gorgeous pink-and-white miniature jewel with grey plastic wheels, complete down to its dainty tail fins, its silver grille and rear lights picked out in red. Satisfyingly, it also boasted a proud innovation, being the very first Matchbox issue with windows, in this case with a cool green tint that suited the stylish, Californian aspirations of the full-sized vehicle. I had one of these back in 1978, picked up from a junk shop for a few pennies, and I liked it very much indeed, without ever realising I was clutching a true piece of Lesney history.

The date was June 1960, and the T-Bird had the honour of capping the Matchbox series of small die-cast vehicles at 75, creating the 1–75 series that collectors have obsessed over ever since. At that time many of the models in the series that had begun in 1953 had been replaced at least once, as Lesney reviewed and renewed its wide offering of cars, vans, lorries, tractors, tanks, buses and construction equipment continually. There was plenty of design refinement, too, led by a change from basic die-cast wheels to smoother-running plastic ones in 1958. Meanwhile, licence was taken with the 'matchbox'-sized boxes, which were gradually creeping up in volume as Jack Odell's design team edged their models up in size for better detail and perceived value.

Since the 1956 introduction of the Models of Yesteryear veteran and vintage series, the company had been busy introducing new lines. In 1957 came the Accessories range, mostly consisting of items that could enhance the enjoyment of 1–75 vehicles such as garages, petrol pumps and road signs, plus a Bedford car transporter. Also in 1957 came the Major Packs, larger models initially in the same scale ballpark as the 1–75s, and that included several very attractive articulated commercial vehicles. These would lead on to the introduction of Matchbox King Size – a line-up, at first, of larger scale road-making equipment that would branch out into bigger lorries and even cars that started to munch directly into the markets of the bigger Corgi and Dinky Toys.

Lesney, as you can see, was experiencing massive growth and with this success came unexpected problems. In the late 1950s, British companies were forced to pay a punishing surtax on their profits, but Lesney's accountants calculated this could be avoided if the company went public and joined the London Stock Exchange.

The sticking point for Smith and Odell, however, was the running sore of the mistrustful relationship with their distributor Richard Kohnstam of J. Kohnstam Ltd. For a flotation to work, they couldn't have an 'enemy' owning half their precious trademark.

They already had disagreements with him over export market strategies and these made the partners doubly determined to jettison him. They didn't have to wait for long because Kohnstam, sensing the relationship was getting too fractious to be repaired, surprised them by deciding not to do battle any longer and in 1959 he sold his entire company to Lesney for £80,000. That would have been a substantial sum at the time – £1.3 million in today's money – and Kohnstam probably thought it was a good time to bail out, rather than end up ultimately facing Smith and Odell in a courtroom. He could have bought himself a brand-new Aston Martin DB4 for £3,980, while barely denting his new-found fortune.

The stage was now set for Lesney Products & Co. Ltd to make its public offering. The company earned 80 per cent of its income from making toys, of which half went for export, and 20 per cent from light industrial die-casting – items like car door handles. In September 1960, 400,000 25p ordinary shares went on sale at £1 each. It was oversubscribed ten times, with 12,415 applications for more than 6 million shares. Within four years, those shareholders smart enough to have held on to their stakes saw their investment values increase fivefold. Profits rose from £425,000 in 1961 to £750,000 by 1963 and had soared to £2.26 million by 1966. In about fifteen years, two dabbling blokes working from a pub basement had transformed themselves into Europe's fourth largest toymakers.

Odell achieved his ambition to become a millionaire by the age of 40. 'I made it with a few days to spare,' he chuckled in later life. Nonetheless, he still regarded the Monday morning new-product meeting as the most important time of each week, where he led the discussions among half a dozen key managers about what new Matchbox models might be made next; the range was now so huge that a new release was called for, on average, every fortnight. That was just the start of the design process. Mock-ups made from Perspex and brass, once approved, would be handed to the drawing office where detailed plans would be made. These were translated into an accurate master model at up to four times the actual size. It would then be split in two to create a so-called 'cataform' from which resin casts would be taken, and then the delicate stylus of a pantograph would reduce each component down to one-quarter scale – Matchbox car size – in chrome-vanadium steel to create the die. Many checks and tests would be run before the mould itself was then struck and hardened, ready for its life of turning out millions of identical toy car parts.

The 1–75 series was the bedrock of the company's success. Taking a leaf out of Corgi's book, the models adopted more and more features that couldn't fail to delight their new junior owners. By 1963, the Matchbox Mercedes-Benz 220SE, for instance, had opening doors, a detailed interior, spring suspension and front wheels that could be steered. Miniature Ferrari, Aston Martin and Maserati racing cars, and motorbikes with sidecars, had beautifully designed spoked wire wheels. Lorries came with plastic loads of huts, scaffolding, girders or pipes. And buses had authentic stickers as livery, which now replaced transfers after Odell had declared: 'Solvent decals make the factory a bloody mess.'

The attention paid to adding tiny details that would enliven the junior imagination reached new heights. The Commer Ice Cream Van didn't just have a realistic Wall's livery but there was a vendor inside, holding a 99; the MG 1100 had a dog on its back seat, sniffing the air through an open window as the car (in your imagination) rushed along; the Pony Trailer came with a drop-down ramp and two white horses. Just one or two of these could keep a youngster happy for a long time, but, of course, many children were accumulating fleets of Matchbox models, and playing interactively with them all could provide hour upon hour of entertainment. Lesney obligingly made forty-eight-vehicle vinyl collectors' carrying cases that encouraged more and more purchases.

For all their run-ins with Richard Kohnstam, Smith and Odell would have one reason to be grateful to him. One of his contacts in the USA, Fred Bronner, became the Matchbox distributor there in 1956. Initial sales were so strong on the other side of the Atlantic that everyone working at Lesney in London received a bonus. In 1959, this New York super-salesman shifted 4 million units in the US – they sold for 49 cents each. Five years later, Lesney made Bronner an offer he couldn't refuse and paid him £261,000 to sell out his Fred Bronner Corporation to them, which then became a Lesney subsidiary with its founder as president. Now the partners had total control of their destiny in the world's biggest toy market.

It was easy to see how the little cars and trucks made such irresistible impulse buys. Yet the other factor in the wild success of Lesney's Matchbox line was that, until the early/mid 1960s, there was no other small-scale range to rival it. No one else yet made little die-cast vehicles like Matchbox and the brand name itself was soon as generic a term as Hoover, Sellotape or Tannoy.

To meet the escalating demand, Lesney was turning out a million toy vehicles a week, which it pointed out cheekily was more than the combined total of the global 'full-size' motor industry. In 1962, Lesney established its own metals division, the Eastway Alloy Zinc Co. Ltd, capable of processing 600 tons of zinc every week, of which 350 tons was sold to other non-toy manufacturers. An in-house plastics department generated 50 million tiny parts weekly. And to provide an idea of the huge expansion that the 1960s held in store for the company, by 1969 its output had risen to a million units a day.

Smith and Odell continued to be extraordinarily capable joint managing directors. Leslie Smith, in particular, seemed to be a business genius, deftly handling each step in the company's expansion. By 1959 Lesney occupied most of the existing buildings in Eastway on the Hackney Marshes and after acquiring a vast site next to the canal on Lee Conservancy Road nearby, the company proceeded to construct a gigantic new factory – its first purpose-built premises – whose imposing concrete edifice became an instant local landmark when it was completed in 1963. This enormous plant increased output by 125 per cent and was home to more than 750 pieces of custom-built (by Lesney) machinery. By 1966, the workforce stood at 3,600, and Lesney was easily the largest employer in the London borough of Hackney, as well as being one of the biggest post-war industrial start-ups in the whole of Greater London.

So that his firm could meet the demand for Matchbox, Yesteryear and King Size die-cast vehicles, Leslie Smith needed labour, and he found it in the armies of local women who worked full- and part-time on the assembly lines. The pay was competitive and Leslie came up with a novel brainwave to make sure the Matchbox conveyor belt would never be wanting for nimble assembly hands. He bought up a fleet of fourteen redundant buses from London Transport, mostly AEC Regent RTs, and had them repainted in the distinctive blue and yellow of the familiar Matchbox toy packaging. These offered free transport to and from work and to assist with the school run for employees' children on route. Perhaps not surprisingly with this perk, there was never any shortage of production-line staff despite the fact that such tasks as assembling the vehicles, applying the stickers and packing the finished products into their famous little boxes can't have been anything but deadly dull toil. It would be the kind of factory work upon which sociologists would make studies that drew grim conclusions about motivation and frame of mind. And yet, it seems, camaraderie, a good working atmosphere and the plentiful availability of flexible hours seemed to prevail at Lesney. As Len Mills, a former tooling and engineering manager, recalled: 'It was like a family. Lesney was not so much a job as a way of life.'

Alan Anderson recalled how his mother Mary took a part-time job at the local Lesney plant in the mid-1960s, setting off for work each night after he'd come home from school and had his tea:

> I think it broke up the drudgery of being at home, plus it helped to fund a new car for the family, a 1963 Vauxhall Victor, and a brand new Pemberton static caravan. She earned something like £6–10 a shift, I'm sure the extra wage also helped us cope when my father, a dock worker, was on a six-week strike in sympathy with Canadian dockers in 1967.

Mrs Anderson was one of hundreds of Lesney employees, overwhelmingly female, to make good use of the free bus travel.

Working from around 5–9.30pm, her job was in effect quality control and she worked in what they called the Fettling Room. Along with other women, she would sit by a conveyor belt full of bare model bodyshells and check them all for major manufacturing faults before they were passed and sent for painting and assembly. Dud ones were binned but quite a few fell into handbags, quite innocently, I'd imagine, as they weren't much good for playing with! Often, however, fully-built cars found their way there, too. They could be one-off experimental ones in different colours to what were ultimately marketed. I think my nephews have them now, in a tailor-made Matchbox storage case. I shudder to think their worth due to sheer rarity if my brother and I hadn't played with them!

Alan, of course, knew the truth but few other keen Matchbox collectors had any idea: the die-cast toys that gripped their imagination and sucked away their pocket money were actually made by other boys' mothers!

To a newcomer, this female realm could be intimidating. A 17-year old Gerry Robinson – the future boss of Grand Metropolitan and Granada – arrived as an inexperienced cost-control junior in the accounts department in 1965. He realised he'd had something of a sheltered upbringing, educated by the Holy Ghost Fathers at Castlehead's St Mary's College in Lancashire, the moment the young man walked through the plant. 'It was wonderful and terrifying,' he told *The Independent* in 2011. Nonetheless, it was a great place for an ambitious young fellow to thrive. He was encouraged to study accounting, something he had never considered before, but also to fully understand the way businesses should be run.

'The beauty of Lesney was it didn't have layer on layer of management. [My manager] Bob Tanner was very clear, and good at making you think for yourself, giving you a go at doing things, and he appreciated the effort.' He also got to understand, and enjoy, the way workers could take advantage of the lack of supervision. Robinson revealed to author Ruth Tait how he would go into the Hackney factory on Saturday mornings to work overtime but after a couple of hours would take an unauthorised ninety-minute break for sandwiches and cards ... until he got caught. During his time at Lesney he progressed through various accounting roles to become chief management accountant, and he only left the company in 1974 because he wanted the company car that Lesney could not provide.

In 1967, in no small part thanks to the contributions of ladies like Mrs Anderson, Lesney was accorded its place in *The Guinness Book of Records* when its annual output reached 100 million models. Three-quarters of them went abroad, with the USA and Japan the two leading destinations. So, there was a blizzard of jaw-dropping statistics to bandy around throughout the 1960s. But what about the products? That was Odell's domain.

Although he had had no formal training, he was in fact a natural manufacturer. Not only was he in overall charge of the creation and manufacture of Matchbox vehicles, but he also designed and built most of the machinery Lesney used to make them. He corralled a large team of talented designers and skilled toolmakers, and trained many apprentices in the company's own way of doing things. One clattering gadget Odell oversaw, for example, was an ingenious machine that included two silos, one of shiny metal wheel hubs and the other of moulded black plastic tyres, the apparatus cladding the wheels with the tyres – something that would have been impossible to do with human fingers because they were so fiddly – and enabling this realistic little feature to grace toy cars sold at pocket-money prices. It could turn out 7,000 completed tyre/wheel units per minute. Odell's other machines pressure sprayed tiny silver headlamps on the cars, or moulded one-piece plastic interiors. It was equipment such as this that enabled Matchbox, at its peak, to be turning out 1,000 model vehicles a minute. 'I just happen,' he once explained, 'to be a born engineer.'

Odell approved and oversaw the design of everything Matchbox. Like Marcel van Cleemput at Corgi, his team worked from manufacturers' drawings and photographs, and their own data, and Odell – who was almost constantly on the factory floor in his white coat, his moustache bristling – had a similarly sharp eye for detail. On one occasion, he intervened in the work of a young engineer wrestling with a prototype for a Ford Model T in the Yesteryear range. Odell felt the driver's seat, measuring around a centimetre wide, didn't look padded enough to be realistic. With no office or secretary of his own, the restless Odell was omnipresent. Yet he had an easy lathe-side manner that encouraged productivity; he sought to discourage traditional hierarchies, lunching unostentatiously in the staff canteen (albeit at an executive table).

With Leslie Smith's agreement, Odell would press the button on each new model, running off an initial batch of 500,000, and then organising the manufacturing roster to produce more if it went down well. In the late 1960s, in particular, the content of the expanding King Size range reflected Odell's admiration for impressive lorries and the intricate workings of construction and agricultural equipment – subject matter that no doubt created plenty of engineering challenges and headaches but were sure to delight the kids who played with them.

Almost everything the duo launched was a runaway success. A few of the products they created, though, weren't winners. One of them was Matchbox Motorway, a cut-price competitor to Scalextric and other electric slot-car racing sets. The crafty bit was that you could use your existing Matchbox die-cast cars on it. Alan Anderson was one of many late 1960s kids who had a set:

> It ran on coiled spring wire which you had to feed into the track once you'd assembled it. Powering the springs were innovative motor boxes – made up to look like racing pits – using a transformer, and their nylon teeth meshed with the springs to power the track. You attached the cars to the coil spring by way of plastic pins which were stuck to the underside of the models. This meant that you had a huge choice of 'racers' to pick from or you could also run convoys of vehicles. A trick my brother found was not to fit the pins in the middle of the model [base] as suggested, but one at the front and another at the back so it gripped like it was on rails. You could go flat out and your car rarely fell off the track! To be fair, it was never as good as Scalextric and those springs were incredibly noisy in the plastic track grooves, but my mum got it at a generous discount – and it was better than nothing!

Lesney's contribution to the UK economy was such that the government felt the need to show its appreciation. Labour prime minister Harold Wilson had urged industrialists to harness the 'white heat of technology' and there was no doubt Smith and Odell had done just that to spectacular effect. The company was awarded its first Queen's Award for Export in 1966, with two more following in 1968 and 1969. Meanwhile, in 1968, Leslie Smith and Jack Odell were both bestowed with the Order of the British Empire.

In that year, the company achieved cultural immortality. America's Smithsonian Institute declared its toy cars as 'perfect examples of their type' and bundled up a selection of twenty-one in a time capsule buried in Amarillo, Texas, for fascinated examination by future generations. Sales of £28 million to some 130 countries generated an excellent £5 million profit in 1968, and at the start of the following year Lesney employed 6,000 people in fourteen London factories. Documents held by The National Archives show that in 1967–68 the company was exporting 79 per cent of its output annually in this peak period – the industry average was 28 per cent. Lesney truly set a blistering new pace for Britain's toy-making sector.

Leslie Smith OBE produced a hardback children's book in the style of an annual and entitled *Mike and the Modelmakers*. Illustrated by Czech artist Miroslav Sasek (his famous children's books included *This is London* and *This is New York*), it told the story of an American boy brought to Britain by his father to show him the 'world's largest car maker': Lesney. The company was on top of the world. The attractive book relays some wonderful facts about Matchbox production in the late 1960s, such as that the mild-steel axles required daily consumed 60 miles of wire; that old biscuit tins were used to carry up to 200 die-cast car bodies each around the factory between production processes, because they were so light; and that the 1-ton zinc ingots used as die-casting raw material arrived at Lesney's canal-side foundry by barge from London's docks.

Jack Odell OBE hit his fiftieth birthday as an extremely wealthy man, with a personal fortune of £35 million. He did not disdain the trappings of success, either. He'd built himself a ten-bedroom mansion at Totteridge, north London, whose swimming pool had a glass-fibre dome so it could be used all year round, sitting in an acre of grounds. In 1969 a false rumour of Odell's death sent Lesney's shares plummeting. As he came off the eighteenth green at his golf club, he said that someone had obviously wanted to knock down the price but added that he'd decided to take no further action, beyond having a pint at the nineteenth to forget about it.

8

Designed in Northampton, Built in Swansea: Corgi Toys Fed the Imagination

For Corgi Toys, the 1960s was one long party – soaring sales, booming prosperity, a shower of awards and products that led the world for the unprecedented ingenuity poured into their creation and the excellence of their manufacture. It was a British success story through and through, and its legacy for collectors and, indeed, car enthusiasts in general is unique.

Right at the centre of this was the brand's star-spangled partnership with James Bond and Aston Martin (see Chapter 14). But Corgi's modelling brilliance and uncanny ability to seed desire for its wares in both children and adults exploded in every direction. In the closing months of 1959, there were early indications of the obsessive attention to detail that would bring Mettoy such incredible rewards.

Hitherto, the company had detected encouraging fillips in demand when it introduced minor innovations. The versatility of plastic certainly helped, such as the Nylon sliding plastic shutter in the side of the Karrier Bantam Lucozade van in August 1958 or the first detailed interior in the Plymouth Sports Suburban in July 1959. This interior cabin detail was quite a simple thing – a flimsy strip of coloured acetate vacuum-stretched over a mould that produced the shapes of the seats and the dashboard, and into that dashboard a contrasting polythene steering wheel was then inserted through a pre-cut hole. Sealed inside the car, this was amazingly effective when viewed admiring through the plastic window glazing section also installed behind the screen pillars. In October 1959, however, Corgi hit its stride in making a model 'feature-packed'. The car was the Renault Floride from France and, as well as the detailed interior, it premiered a new suspension system entitled Glidamatic that made the wheels springy and enhanced the realism of the pretty lines of the original car as it was played with. Sales would eventually reach just under half a million units.

Two months later, in December 1959, Corgi launched a Chevrolet Impala Police Car with the same cocktail of realistic attributes as the Renault but with extra touches: a 'State Patrol' sticker on each door and a tiny Nylon plastic radio antenna on the rear wing. This little package sent demand stratospheric. On sale for five years, this was the first Corgi model to sell more than 1 million copies (1,073,000 to be precise). It was in the vanguard of Corgi's sales success in the USA, where annual turnover increased by a third in 1959, as total export sales rocketed by 41 per cent and the range also entered the Japanese market. In August 1960, Corgi pulled off a similar trick with a British motorway patrol car based on the Ford Zephyr MkII estate, which proceeded to sell an extremely healthy 771,000 units.

Before that, in March, Corgi had gone even further, shocking its competitors and delighting its customers: the Aston Martin DB4 had an opening bonnet with Aston's straight-six engine intricately detailed in cast metal beneath it. That was in addition to interior, suspension and wheels detailed to look like the real car's. 'This is indeed the biggest triumph yet achieved by any die-cast scale model car manufacturer and the climax to the rapid, yet successful, series of new and revolutionary ideas brought out regularly from the Corgi workshops,' said a breathless newsletter sent to stockists to gee them up. Product and hyperbole combined worked a treat – Corgi sold 982,000 examples over five years.

Van Cleemput and his team were really motoring and in April 1961 they produced a model car that was even more desirable than the Aston, again with a world-first mixture of features. Corgi's Bentley SII Continental was a superb evocation of the original with its elegantly balanced two-door bodywork by H.J. Mulliner. Up close, it was stunning, and not just in its chrome-plated radiator, bumpers and Bentley mascot. It had 'jewelled' head- and tail-lights (the reflective glass beads were supplied in enormous quantities for this and later vehicles by Birmingham's Rhinestone Enterprises) that gave the two-tone Bentley an extra precious quality. At the front was an ingenious self-centring steering system, its intricate mechanism housed unobtrusively in the car's nose, while at the back the boot lid opened to reveal a removable spare tyre/wheel in its own storage nacelle on the boot floor.

At 964,000 sold, it was another of Corgi's near-million sellers, and as van Cleemput recalled in his definitive Corgi book, his boss Arthur Katz was particularly proud of it. Corgi was already exporting big time to more than 100 countries when, in April 1961, Katz and his wife set off on a two-month tour of the USA, Hong Kong, New Zealand and Australia in a sales drive, and he regarded this Bentley as his best visiting card. He was back, incidentally, in time to travel to Buckingham Palace in November where the Queen bestowed the Corgi managing director with an OBE for services to the toy industry.

This beautiful object, a fine cocktail of the modeller's art and the mystique of great toy-making, was important to two newcomers who hopped aboard the Corgi express train early on. One was 16-year-old local art student Tim Richards, whose talent had been cannily spotted by Mettoy in Northampton. In a 2017 interview with the author, Tim said:

I went to art school and I'd done loads of local exhibitions of my sculptures, and Henry Ullmann kept offering me a job, I kept saying no, sorry, I'm an artist. You know, it was 1959 and I was all black polo-necks and French cigarettes! But eventually I succumbed because they offered me a great career opportunity plus art school outplacement. I was very good with Plasticine so suddenly I was employed making the wax originals for squeezy baby toys, Walt Disney's Lady & The Tramp, things like that. Shortly afterwards, Mettoy switched to making vinyl footballs, and those didn't need any sculpting, so they shuffled me into the design department of Corgi. Marcel [van Cleemput] was there. He was definitely a self-promoting individual and any compliments went to him – we were never in receipt of any congratulations. He wanted to be known as the genius behind Corgi, and he was, but not to the extent he claimed. However, he very much wanted to be part of our team, to go to the pub with us, which naturally we weren't too keen on. I was very good with modelling materials so I was basically a paid sculptor, and we all had our pigeonholes. A lot of the team were ex-Bassett Lowke [model trains and locomotives] and so they'd made the transition from artisans to mass production.

As an artist working in industry, Tim's talents found many outlets. Back to the Bentley Continental, usually issued in a gleaming, stove-enamelled black-over-silver:

When a new model was coming out, we were often asked to paint them for toy fairs. I painted a Bentley Continental olive green and ivory because I thought it was a great combination, and it pleased me. I was amazed when they picked it up and loved it, and that version went into production.

His speciality, though, was figure modelling and among many examples of his work to be found across the Corgi canon, this was typified by the plastic evocations of the actors and wild animals that brought alive Corgi's Daktari range from the popular TV series. Tim stayed with Mettoy until 1972, leaving after the Corgi Magic Roundabout series was issued containing the ultimate examples of his work for the company in the form of plastic figures of Dougal, Ermintrude and the rest of the characters.

Meanwhile, experienced patternmaker John Marshall – then designing tooling for Kitmaster plastic OO-gauge model railways at nearby Wellingborough – had admired the Bentley's attractive attributes when he saw it in a toyshop, and jumped at the chance to join the Corgi backroom boys in Mettoy's 'sample room' – the internal term for the prototype or experimental department – in 1961. The twelve-strong team had very diverse backgrounds, John said, including four who had been there since the earliest days of Mettoy, and the manager Percy Wilford had gleaned much of his casting knowledge from his time as a dental technician.

'My first job was the plastic top for a Corgi articulated horsebox,' he remembered in 2017:

> In the end it would be made from Styrene with MEK, the liquid solvent that welds the plastic together instead of using plastic glue ... Marcel van Cleemput and the directors would collectively decide what to make, and then a first mock-up would be created from drawings and photos. Then a wooden patternmaker would carve it at 1:11 scale – four times the size of the final model – from a piece of lime wood that was quite hard to get; we had a permanent order with a local timber yard in case they ever got hold of a felled lime.
>
> Then a pantograph would trace the wooden version down to the scale we needed in resin, which would accurately capture and reduce the detail. When it came to the individual components, we would give the drawing office next door an idea of what we wanted, they would draw it, then someone in the sample room would make that exactly to the drawing, to prove the drawing. And this went on year in, year out.

Just as for Tim, John found working with Corgi allowed plenty of opportunity to shine:

I loved experimenting, I designed a lorry with a working tail-lift in 1962 and although they never made it, it got the attention of Philipp Ullmann [Mettoy founder]. It was a very good grounding for an engineer because you needed to come up with things that were cheap, easy to assemble, and reliable. As there was a huge production run to come, development costs were negligible; we'd be making millions. So you could devise neat ideas and the company would say yes to them. Well, most of the time, anyway; one of mine that didn't get made was working windscreen washers, using modified fountain pen nibs!

Other 1961 Corgi cars with satisfyingly clever features included the Ford Consul Classic, whose bonnet sprang open if the front wheels were pressed down; a Triumph Herald Coupe whose front wings and bonnet pivoted forwards, as on the real thing, to reveal its engine; and the Chevrolet Corvair, whose die-cast engine was under a lifting cover at the back and also had a tiny sunblind inside its rear window. There were multiple issues in an attractive Chipperfields Circus set and a Jeep pick-up with an elevating platform on which a tiny workman could change the bulb in a streetlight. The Ecurie Ecosse Racing Car Transporter thrilled motor-sport enthusiasts with its accurate evocation of the unique actual vehicle, and Corgi's array of tractors and farm equipment took the brand's by now trademark infatuation with realistic detail and working parts to the rural fanbase.

Yet another novelty was Corgi models that were vacuum plated to look as if they were gold plated. The Bentley, Consul Classic and Corvair thus treated, which required quite a lot of hand polishing, went into a gift set called Golden Guinea. Yet another range was the shiny golden Trophy Models of sports and racing cars on little display stands. They were produced exclusively for the Marks & Spencer chain of department stores. M&S had sold Mettoy products in the past and Philipp Ullmann and Simon Marks were close friends. Peter Katz, who later succeeded his father Arthur as Corgi managing director, recalled the relationship:

> Philipp Ullmann lived in Abbey Road, St John's Wood [north London], literally on the other side of the road from the studios, and I think he invited Simon Marks to his flat to examine the new toys that were being put forward. It was a very personal business in those days, even in very large companies.

Throughout 1962 and '63, the Corgi bandwagon rolled on. Van Cleemput's *Great Book of Corgi* records all the many cars and other vehicles that were issued. A few that were enormously popular were the Heinkel Cabin Cruiser bubble car, the Ford Thames Airborne Caravan motorhome, the Jaguar MkX and a Carrimore Car Transporter or Low-Loader refreshed with the latest Bedford TK cab. The Citroën DS19 Safari would appear in numerous versions, while the Ferrari Dino 156 Formula 1 car had its distinctive 'shark-nose' replicated beautifully and delicately in die-cast metal.

It was perhaps natural that van Cleemput would turn his attention to vehicle lighting and he devised a nifty if not especially effective system called Trans-O-Lite that took daylight from the roof of a Volkswagen or Commer van or a Rover 2000 car, and piped it, fibre-optic-style, to the headlights. Another innovation, first seen in 1962, used the first inclusion of an AA-sized battery in a Corgi vehicle; it powered the beacon and roof-lights of a Superior Cadillac Ambulance (Superior was the name of the ambulance conversion company, although the word willingly doubled as a complimentary adjective). As the design work became ever more complex, so the timespan needed to get a model from idea to marketplace stretched from twelve to fourteen months although, as van Cleemput was keen to point out: 'The lead times of some of our competitors were in the region of two years or more.'

And that does bring us neatly to one of the most impressive of 1963 issues – a toy car that, for its 1:48 scale, had more functional and decorative design features than anything else seen up to that time; 'the works' is how van Cleemput referred to it. In November, Corgi's Ghia L6.4 took its bow and set a fantastic new standard in die-cast model engineering. It was the first ever to have opening doors, bonnet and boot together, and while a detailed engine, jewelled headlights and working suspension had appeared in previous Corgis, none of them had tipping seats like this one. There was even a tiny plastic Corgi dog, panting on the rear parcel shelf.

Pretty much no one had even heard of the Ghia itself, a car so exclusive and limited in production that only twenty-six real-life examples were made and sold to such celebrities as Frank Sinatra, Dean Martin and Lucille Ball. Basically, you were never likely to see one outside Beverly Hills and certainly not on the roads of Britain. And yet the Corgi model of it, crammed with gimmicks, was a runaway best-seller, with 1 million sold in the first two years alone, and a total of 1,753,000 in its six-year production life.

The factory at Fforestfach was running red-hot as it struggled to keep pace with demand. From selling 3.9million models in 1960, by the end of 1963 it was at a dizzy 12.2 million.

Between 1961 and '63, the workforce expanded from 1,000 to 1,700 and Mettoy was suddenly one of the biggest toymakers in Britain; it would remain in the top four until the early 1980s. The company's board consisted of chairman Philipp Ullmann, managing director Arthur Katz and directors Henry Ullmann (Philipp's son) and Howard Fairbairn, all of them key members in the Corgi launch, along with works director Adam Heaton and sales director Frank Varnals. They'd been fascinated observers as their upstart rival Lesney had gone public in 1960 and on 5 June 1963 the board followed suit when Mettoy Company Ltd was floated on the London Stock Exchange. The £40 million offer, for 620,000 5s ordinary shares at 15s a share, was oversubscribed ninety-one times as punters betted on a repeat of Lesney's wildly successful listing.

Many members of the Mettoy workforce felt the atmosphere at the company at this point shifted to be more corporate and impersonal. The 'family firm' nature of the relationship between the Ullmann and Katz clans and their employees changed. However, on the occasion of founder Philipp Ullmann's eightieth birthday in 1963, a lavish party was thrown for everyone at the Northampton HQ, and Mr Ullmann was driven up from London, with his private nurse on duty, to attend. 'Even though the Swansea factory was enormous, Northampton was Ullmann's base, and he was very attached to it,' said Tim Richards, who was among the hundreds of guests at the town's Grand Hotel (now a Travelodge) served with champagne and real caviar.

The year 1964 was another bumper one for new Corgi releases packed with surprise and delight. The Corgi Major Bedford Simon Snorkel Fire Engine was an icon among all 1960s toys, with its hand-operated double boom and a platform that was kept level at all heights by the clever incorporation of a pantograph system. It sold more than 1 million units and was on sale for twelve years. And in November came another big hit, a Mercedes-Benz 600 limousine with windscreen wipers operated by the wheels as the car was pushed along. Both vehicles had Glidamatic suspension and beautiful detail finishing that belied the technical wizardry inside them.

The issue of spying started to be taken very seriously in Northampton, and not just in the form of tiny plastic detectives and special agents. 'There was definitely industrial espionage,' claimed John Marshall. 'There was a period when every time we did a model, Dinky did the same one at the same time, such as the Routemaster bus.' The first floor at Harlestone Road was soon like Fort Knox. The drawing office and sample room were identical spaces, 30ft wide by 100ft long, and no unauthorised person was allowed in either, such was the febrile competition gripping Britain's toy car industry in the mid 1960s.

Superb and standard-setting detail was intrinsic to another important 1964 unveiling: Corgi Classics. This range of models of veteran and vintage cars was timed to commemorate the thirtieth anniversary of Mettoy. The commercial imperative, though, was to nab some of the market Lesney had created with its nostalgic Models of Yesteryear. Jack Odell over in east London may have prided himself on the fine cast detail in his Yesteryears and the particular care he lavished on that range as he oversaw their design personally. However, van Cleemput was determined to do better – much better. The project was indeed quite a personal one, as the Frenchman undertook all the design himself and barricaded himself away in his office for three hours each day for weeks to create the dozens of tiny components – including the fastidiously intricate wheels with their astonishingly thin spokes – for the 1915 Ford Model T and 1927 Bentley 3-litre Le Mans car. The results were truly superb and actually made Yesteryears seem rudimentary by comparison.

Mettoy held a bash at London's Dorchester Hotel to launch the Classics and celebrate its birthday, where the guest of honour was S.C.H. 'Sammy' Davis, a vintage car guru and writer who also happened to have raced a Bentley in the 1927 Le Mans 24-Hour race. Oddly, though, the Classics weren't especially popular and only a handful of additions to the line would be added before they were discreetly axed in 1969.

There was another development in 1964 that increased Mettoy's output hugely beyond the 13 million Corgi vehicles it would sell that year. Its long-term customer Marks & Spencer had decided to abandon toy sales altogether, which was quite a blow. On another front, Lesney's King Size range was providing increasingly stiff retail competition to Corgi. Now Corgi would hit back at the Matchbox 1–75 series and bag an important new outlet. It developed a range of small, Matchbox-sized cars for a completely new brand, which continued the dog theme by adopting the Husky name.

'We sold them in Britain, Germany and the United States exclusively through Woolworth's,' Peter Katz, who was then a Corgi regional manager covering first Scandinavia and then the USA, explained to the Museum Of Childhood. 'Your generation won't realise how important Woolworth's were. They had between 3–4,000 stores in the US, in Britain they had over 1,000 stores. Germany – I'm not quite sure – 200, which had been regrown after the Second World War.'

The little Husky cars and lorries had cheap plastic bases and wheels to keep prices and shipping weight down, and unlike their Matchbox 1–75 equivalents they came in blister packs that hung on revolving shop carousels. Van Cleemput barely refers to them in his Corgi 'bible', but by 1965 Mettoy had 250 people working on the Husky range in production on site there. There were one or two interesting subjects – Sunbeam Alpine, Lancia Flaminia, Jaguar E-type and Ferrari 250LM – with appeal to car enthusiasts. They were rather brittle and, today, they're not very collectable unless they're still in their blister packs, which, obviously, is very rarely encountered because they were generally torn open on the spot by children of parents with precious little money to devote to toys.

The really big news for 1965 is dealt with in this book in Chapter 14 – Corgi's move into the film and TV world. Other issues were a range of military vehicles, racing cars including Monte Carlo Rally competitors and the Le Mans Ferrari 250LM, a Ford Thames Wall's ice cream van and a magnificent Ford H Series Tilt Cab articulated lorry. There was also a particularly attractive Lotus Elan with a 'Put a tiger in your tank' Esso decal on its boot. Annual sales rose to 14.7 million items. In 1966 (which saw a whopping 16.9 million sales) came some memorable gift sets including the Lotus Racing Team Set, a revamped Carrimore Car Transporter with six cars as one package and a Ford Tractor with Animal Trailer and plastic livestock. One notable car issue was a Mini Cooper S in the livery of the vehicle that Timo Makinen and Paul Easter drove on the Monte Carlo Rally. Breaking all previous records for the speed of new idea to retailer, Corgi managed to get this one into the shops in a little more than two weeks after the event. However, although the

Cooper won the rally, it was subsequently disqualified for a rule-breaking headlight ... so Corgi's celebratory limited edition made do with Makinen and Easter's autographs screen-printed on its roof in place of anything suggesting victory! There were a few other car issues that year too of which I'm especially fond, including the Ford Cortina Mk I estate, Marcos 1800, Citroën DS Le Dandy and a Volkswagen breakdown truck – all of which I would dearly like to own even today.

Anthony Fleischmann, son of Werner Fleischmann whose company Reeves International was the US Corgi importer, recalled this booming period with wonder:

Really, I was just a little boy and my childhood was spent running around the warehouse that my father had in New Jersey. It was absolutely filled with Corgi die-casts, Britains metal figures, Märklin electric trains and Steiff bears. Those are the four pillar brands he started with and distributed for decades over the entire United States. I do vaguely remember the presence of the new generation of the founding families, Henry Ullmann and Peter Katz. I think their fathers were stepping back at this point. But Howard Fairbairn was very important still. I remember my father speaking about him at the dinner table as the genius at Mettoy, the innovator. They were tremendous innovators – it was all about design and innovation, and Corgi was just top notch at it. Their amazing technical experts built in, designed in, engineered in all those incredible features to make that happen. And so the Corgi brand made an awful lot of people very happy.

A full set of Hornby's pioneering 22 Series Modelled Miniatures, ancestors to Dinky Toys, as they would have been supplied in 1933. In August 2012, this ultra-rare and original set was sold for £13,000. (SAS Auctions)

This Dinky Toys gift set of commercial vehicles was produced exclusively for the US market in the 1940s as exports grew; it includes, bottom left, the Bentley Ambulance, which drew on construction methods copied from Tootsietoys.

The incredibly rare Bentalls department store livery on this Dinky 28 Series delivery van – nirvana to serious collectors – gave it a final auction value of £12,650. Dinky was quick to harness the pulling power of other companies' branding. (Christie's)

DINKY TOYS
& DINKY SUPERTOYS

AUGUST 1957

HEINZ
57 VARIETIES

DUNLOP *The World's Master Tyres*

DINKY TOYS

OSBORNE'S
118 High Street
Rushden
Tel. 2415

2d
U.K.

Cover of the 1957 Dinky Toys catalogue features a galaxy of models from the range, jostling for road space in London's notorious traffic blackspot Piccadilly Circus. The supplying toyshop Osborne's of Rushden, Northamptonshire, was still providing junior delight in 2021.

Lesney's miniature world by 1959 comprised seventy-five small Matchbox models, still sold under the Moko brand of its marketing partner J. Kohnstam, with Accessory Packs to complement them, and the beautifully detailed Models of Yesteryear such as this Fowler Showman's Engine, a Jack Odell masterpiece.

Every Matchbox model, just as for rivals Dinky and Corgi, was hand-built by someone else's mum, aunt, grandma or big sister. These ladies are hard at work at Lesney in the early 1960s on the Major Pack production line where countless No. K4 Fruehauf Hopper Trucks are coming together.

Within little more than two years, Mettoy's Corgi Toys range had exploded with cars, van, lorries and military equipment, as the cover of this 1958 catalogue amply shows. If you liked authenticity and detail then there was always something new to feed the attention.

It was a big year for Corgi, 1965, as it approached the peak of its modelling powers. That year's catalogue cover shows the Ghia L6.4 on which everything opened and the 1927 Bentley 3-litre whose exquisite detail heralded the new Corgi Classics line-up.

Dinky Toys worked with MG to ensure that this model – the first ever with opening doors – went on sale on the same day as the actual MGB car in 1962.

The era of die-cast character merchandise began with Corgi's version of its Volvo P1800 with Roger Moore's The Saint at the wheel and in stickman form on the bonnet; it was suggested by the firm's Swedish distributor, and 1.2 million examples were sold.

Tri-ang's Spot On models boasted exceptional detail, these lorries being part of a range that adhered strictly to a 1:42 scale. They're hugely sought-after today, even though their 1959 introduction barely troubled the die-cast 'big three'. (Julie Sherriff)

Blockbuster: the world went crazy for Corgi's Aston Martin DB5 from the hit James Bond film Goldfinger, *in fact a model produced in a rush after early wavering – the clumsy moulding line around the front shown here hints at what its designer called a 'botch job' ...*

Holidays abroad by motor coach

DINKY TOYS No. 953—CONTINENTAL TOURING COACH

Paris, Rome, Brussels, Vienna . . . all these cities, and many more, are visited every year by hundreds of coaches similar to the prototype of this Dinky Supertoy. The model itself is fitted with specially tinted windows, seats and steering wheel and is finished in a delightful pale blue gloss with white roof. The words "Dinky Continental Tours" appear on each side. Length 8¾ in.   U.K. Price 13/6

The last word in realism!

DINKY TOYS No. 277
SUPERIOR CRITERION AMBULANCE
WITH FLASHING LIGHT

With siren wailing and roof-light flashing the ambulance rushes its patient to hospital! This striking miniature also is fitted with a red roof-light which actually flashes as the vehicle is pushed along. Realistic figures of driver and attendant occupy the front seats. These refinements are additional to the regular Dinky Toys features Prestomatic steering, windows, suspension and seats.
Length 5 in.   U.K. Price 8/9

Battery No. 036 (Vidor V16) is not supplied with the model, but may be purchased separately.   Price **5d.**

DINKY TOYS

MADE BY MECCANO LTD.

AVAILABLE LATER OVERSEAS

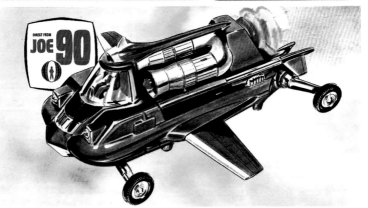

A 1969 ad from a comic finds Dinky Toys creating a furore in the playground with its flying car from TV's Joe 90; this and other vehicles and craft from Gerry Anderson's small-screen extravaganzas, especially the Rolls-Royce FAB 1 from Thunderbirds, *proved a Godsend for the prospects of Meccano's Liverpool factory.*

Left: *Fanfare in the* Meccano Magazine *for 1962 Dinky releases: the Cadillac Superior ambulance with battery-powered flashing beacon aimed for realism but the Continental Touring Coach was plain wrong – it was Dinky's American Wayne school bus in a new livery, which would never be seen travelling through Europe!*

Corgi's 1966 Batmobile became, over its very long life, the best-selling single item in the brand's history but, as this advert shows, the range was already brimming with incredible diversity, a measure of the dedication lavished on Corgi by its French-born design guru Marcel van Cleemput.

To encourage collecting, Lesney hit on the idea of suitcase-like Collector's Carry Cases, this early one featuring the Matchbox No. 41 Ford GT40 as décor to the lid, below which forty-eight models could be housed. Marketing genius like this propelled it to producing 1 million small vehicles a day by 1969.

Lesney was eventually forced to diversify in the 1970s, but even before that it tried everything it could to spread interest in its die-cast cars and trucks, including a range of small jigsaws featuring popular issues like this Matchbox Volkswagen Camper.

Now the final process begins. All the parts of the model are brought together at the Assembly lines and put into their correct positions at the right time by thousands of nimble fingers. Each person doing their own part in building the model just as they do in the real car factories. Only here, eleven cars a second roll off the assembly lines!

A spread from Mike & The Modelmakers, a book commissioned by Lesney in 1970 to relay the Matchbox phenomenon from Czech author and illustrator Miroslav Sasek; his graphic representation of the vast, bustling Hackney factory brilliantly explained how this No. 21 Foden Cement Mixer came together.

Vivid cover of the July 1970 edition of Meccano Magazine with a photographic array of everything that awaited young car lovers in the Dinky Toys world. The publication described new issues each month with breathless excitement, thanks to long-serving editor Chris Jelley.

The audacious launch of Mattel's Hot Wheels in 1968 changed the toy car world forever; it brought with it a celebration of American West Coast car culture, custom cars and hot rods such as this Hot Heap from the initial, sixteen-strong line-up, supplied in eye-catching blister packs complete with a tin 'collector's button'.

The low-friction axles and wheels on Hot Wheels cars were intended for high-speed action on track systems that could be assembled for stunts and racing, providing the sorts of thrills and spills previously only offered by Scalextric. It proved a worldwide smash.

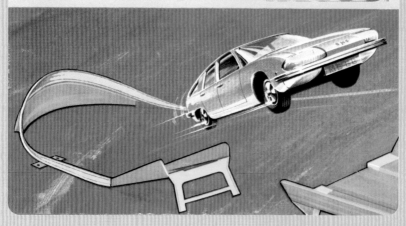

For Lesney the arrival of Hot Wheels proved devastating, with demand and profits blindsided almost instantly; it took the London company eighteen months to re-tool for its Superfast response, including racetracks and a range of new fast cars for them such as this Pininfarina BMC 1800.

Left: *Corgi Rockets was Mettoy's award-winning answer to Hot Wheels: a new range of vehicles like the Tom Sweeney car, illustrated on this Lap Counter box, as part of a Stock Car set came with low-friction wheels, detachable chassis, and little tubes of oil so they could be lubricated to race-ready perfection.*

Below: *Dinky also had to quickly evolve its offerings to square up to Mattel's Hot Wheels, bringing out well-detailed larger cars with low-friction Speedwheels on dream machines like the Lamborghini Marzal; by 1971, however, Dinky's destiny was in the hands of the receivers.*

Early 1970s advert captures three Dinky Toys initiatives intended to pull the brand back from the brink: Shado 2 capitalised on Gerry Anderson's live action sci-fi show UFO; Dinky Kits tapped into the DIY hobby mindset; and cars like the Matra adapted former French tooling to expand the British range.

New owner of Dinky Toys, Airfix, tried to give its die-cast division a new push. This real-life London Transport Routemaster bus was completely repainted as a rolling advertisement for Meccano and Dinky Toys in 1975, and did its best to raise interest on the streets of the capital.

The despatch department at Corgi's Swansea factory warehouse in the late 1970s, with the all-female workforce hard at work trying to meet the demand. Just look and drool at all that desirable mint-and-boxed stock on the shelves, waiting to be shipped.

This Matchbox Models of Yesteryear Talbot van drove into a storm with Buckingham Palace in 1978 over unauthorised use of the royal crest in its Lipton's Tea regalia. It did, however, herald very many more releases resplendent in nostalgic advertising livery.

Team Surtees won the Formula 2 motor-racing championship in 1972, lighting up the grid with the help of Matchbox sponsorship (this is a period sticker); bizarrely, though, there was no Matchbox model of the car itself, although later on there was a Speed Kings Surtees F1 car that didn't have Matchbox livery ...

"Join Crocker and me on law enforcement duties in our Department Buick." With clip-on roof beacon, gun-shot sound and a self adhesive NYPD lieutenant's badge for you to wear.

C290 Kojak• ™ **Buick**

* A trademark of and license by Universal City Studios Inc

C434 Charlie's Angels Van A beautifully decorated Chevrolet Van with a fully detailed interior. *Length 122mm.*

C292 Starsky & Hutch Ford Torino *Length 153mm.* Get into the action with T.V.'s toughest duo. Complete with miniature figures of Starsky and Hutch and a suspect.

In the late 1970s. with James Bond exhausted, Corgi moved its attention to Saturday-night TV action shows, to cash in on the wild popularity of Kojak, Charlie's Angels *and* Starsky & Hutch; *general scale also shifted to a bigger, brighter 1:36 as the marketing battle was set for an onslaught from* Star Wars.

The kind of old-fashioned toyshop found in every British town until the 1980s; in about 1979 A.C. Warren has an original Chopper bike in stock and the very last of the Dinky Toys displayed on the inside windowsill; all that's missing are the brown shop coats and the half-day closing sign ...

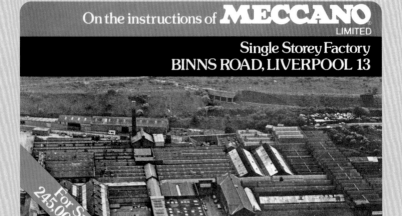

On the instructions of **MECCANO**
LIMITED

Single Storey Factory
BINNS ROAD, LIVERPOOL 13

For Sale
245,000 sq.ft.

Airey Entwistle
Industrial and Commercial Property Consultants

Selling Meccano's factory of dreams – for redevelopment: Airfix closed the Binns Road home of Dinky in November 1979 and turfed out its defiant employees four months later. Once the buildings had been sold it wasn't long before the bulldozers set to.

In parallel with Lesney, Mettoy won the Queen's Award for Export in 1966, 1967 and 1969. In those years, its Corgi export dispatch department was working flat out, shipping 9.1 million models in 1966 and almost 9.8 million in 1968 (the figures for 1969 have unfortunately been lost but would have been slightly lower). Mettoy Playcraft (Sales) Ltd was formed in early 1966 to streamline the marketing of the group's many ranges and as part of that process Corgi's first advertising campaign in the USA was launched in partnership with Reeves. Meanwhile, back in downbeat Swansea, from where all these world-beating toys emanated, the plant was expanded to 385,000sq. ft as the number of employees rose from 2,250 to 3,500 between 1965 and '66. Profits, too, had powered ahead relentlessly every year since the decade began: £110,105 in 1961 to £890,083 in 1966.

Some wonderful commercial vehicles would cause junior noses to be pressed to toyshop windows around the world in 1967. The Ford Holmes Wrecker was a beauty of a tow truck, with its complex chassis and brilliantly designed working towing equipment. There was also stunning chassis detail for the Dodge Kew Fargo as a tipper but also as an animal carrier, which came complete with a haul of pigs on a cardboard load area insert designed to look like straw; it was touches like these that reassured you designers with an eye for fun details were running the Corgi design department – people who could guess what would cause 'surprise and delight' before the end user had even realised it themselves. And who could fail to love the Commer Mobile Camera Van with its cameraman and platform that could be attached to either the roof or behind the rear doors? Certainly not Samuelson Film Service of north London, who joyfully co-operated with Corgi to make the little vehicle look just right. Among new cars, lovers of unusual design touches could choose a Lincoln Continental limousine with a TV screen in the back (yes, really: interchangeable, channel-hopping screens were backlit by a battery-powered light), or there was a Lancia Fulvia Sport Zagato with its bizarre, side-opening bonnet. This last was probably a car van Cleemput had noticed on one of his many visits to motor shows around Europe, where he sought inspiration ... and was unique among his peers in doing so.

Corgi's big deal in 1968 was 'Take-off Wheels' with 'Golden Jacks'. It was an idea John Marshall came up with one night while relaxing on the couch in front of the TV, and Mettoy director Howard Fairbairn eagerly agreed to incorporate it into existing and forthcoming cars. He even halted some existing production lines to get cars so equipped flowing down them in double-quick time. The system featured a pull-down jack in the form of a short metal rod fitted behind each wheel. When they were lowered, the separate wheels could then be removed, allowing you to change wheels just as you would in the real world of motoring. When they were raised, the wheel was locked into place, and the spring suspension and the car's ability to roll along freely were completely unaltered. The Rover 2000 and Oldsmobile Toronado were adapted with the features, but new issues including the Mini Marcos and Chevrolet Camaro were designed around them ... and still managed to cram in opening doors and, on the Marcos, an opening bonnet, and retractable headlights on a particularly gorgeous and covetable Chevrolet Corvette Stingray.

Other extremely pleasing car models that year included a Mini Cooper with a sliding sunroof, a

Renault 16 equipped as a camera car for the Tour de France cycle race and a truly superb Jaguar E-type 2+2 with opening everything and adjustable front and rear seats. Respect must also be paid to the American LaFrance – an astounding model fire engine, executed stunningly and with its turntable escape fully functioning in every which way, that seemed to contain more components than a TV set. It was such a winner it was on sale until 1980, with sales kissing the 700,000 mark. Development of the LaFrance once again benefited from the meticulous approach of initial project leader John Marshall. He built a special test rig to raise and lower its ladder-winding mechanism, to ensure the breaking point would be well beyond the constraints of the delighted play that it encouraged.

This book is not an exhaustive list of everything Corgi issued in the 1960s but I can't let the decade pass without touching on a couple of the great models of 1969 that I've long admired for their technical wizardry, superb finish and the must-have aura that Corgi packaging designers always got dead right in their boxes, displays and illustrations. The way that cars such as the De Tomaso Mangusta and London–Sydney Marathon-winning Hillman Hunter were presented was likely to have a young-ish collector on the horns of a dilemma: play with the car and risk damaging it or keep it in its box to admire but not enjoy the ownership experience to the full? However, there was probably no such conundrum for the very young, pre-school target market for a range of new, simple, heavy and ultra-tough Corgi Qualitoys trucks. These shared an identical chassis but sported different rear bodywork ranging from a fire engine to a horsebox, in much the same manner as the

Tiny Tonka line-up of vehicles. The key difference was the Qualitoys were die-cast while the Tonkas were made (in Canada or Japan) from pressed steel sheet. Although any self-respecting 1960s schoolboy with even a passing interest in the exciting real-life vehicles of the era would have scoffed at them, the reception from the British toy trade was unusually warm. It was a salutary lesson that, no matter how active and free-thinking the imagination of van Cleemput and co, Corgi was in the business of making toys and not pieces of engineering art. Sadly, for all of us whose love of classic cars was stoked by Corgi's epic decade of scale-model excitement, a new and unexpected reality check was just around the corner.

9

Home-Grown Rivals Who Chased the Big Three

Muscling in on Meccano's early stranglehold of the die-cast toy-vehicle market was always going to be tough. Few other companies embraced the relatively new techniques of die-casting in mazak like Frank Hornby's Liverpool enterprise, and none tackled it on such a huge scale. At the same time, Meccano skilfully built up the Dinky Toys brand so that it was a byword for quality and then charged premium prices in the best of retail outlets, such as leading department stores and the largest toyshops in each town.

Tri-ang's Minic series of tinplate and clockwork vehicles did present strong competition, albeit at higher prices, and throughout the 1930s and '40s there were many other tinplate car and lorry alternatives to Dinky Toys at all price levels. One brand to enjoy widespread distribution was Wells Brimtoy, whose range of smaller tinplate lorries with friction-drive matched Dinkys on price, and other familiar names included Mettoy, Chad Valley and the UK factory of

American toy tycoon Louis Marx, plus imports from German firms such as Schuco.

After the Second World War, with so many people having worked and innovated in metallurgy in the defence effort, smaller companies felt emboldened to nibble away at Dinky's dominance with their new-found casting experience. Timpo was a manufacturer of wooden toys that added a few generic die-cast cars and lorries to its range before deciding to stick principally to hollow-cast and later plastic toy soldiers; Crescent was another, producing a handsome line-up of sports and racing cars that were a cut above the Dinky equivalents in detail design but not, alas, in quality, as inferior materials or contaminated machinery would lead to metal fatigue on the cars themselves. They lasted until 1960.

By that time, both Lesney and Mettoy had decided to take Meccano on directly with their respective Matchbox and Corgi ranges. There were a host of other rivals, too. Lesney founder Rodney Smith had provided his die-casting expertise to London toy wholesalers Morris & Stone, overseeing manufacture of both the early Morestone range – similar to Matchbox cars – and being instrumental in transforming it into the Budgie brand, which did a neat range of Trucks of the World and various other vehicles. Some of these were still on sale in the 1980s, buses and taxis sold as souvenirs in London's cheap 'n' cheerful tourist gift trade, but Budgie's modest success never really troubled Britain's Big Three.

A more serious challenge came from Die Cast Machine Tools Ltd, based in north London and then Hertfordshire. This company's efforts were poured into making die-cast toy guns that cashed in on the craze for Western films just before and after the Second

World War under the highly appropriate Lone Star brand name. Exacting manufacturing standards and a healthy export market in the USA made die-cast cars a natural extension and the first Road-Master series of old and new subject matter – an ancient Ford Model T next to the latest Thunderbird sports car – was a toe in the water. Then in 1960, DCMT came up with a range of mostly contemporary American cars at 1:50 scale (not far off the size of most Corgi cars) offering largely what their established rivals did – windows, suspension and chrome plating – but with their short axles denying them decent realism because the wheels and white tyres were tucked too far inside the cars' bodywork. They were marketed using the venerable Tootsietoys brand name in the US and as Roadmasters (with the hyphen dropped) in the UK. These sold quite well but DCMT was now eyeing the rampant success of Matchbox and so in 1965 the company had a cunning rethink, resulting in its Roadmasters Impy series. These were to a smaller 1:60 scale, so almost Matchbox-like in scope, but the designers incorporated multiple features, said to be thirteen per car, just like the larger Corgi and Dinky Toys. That meant opening panels – one, a Peugeot 404, had a rather pleasing open sunroof – detailed interiors, windows, suspension and even jewelled headlights. It was a bold undertaking and the range did reasonably well for a few years, although it was definitely an also-ran to Matchbox's all-conquering 1–75 Series. The final Lone Star assault on the toy car world came in 1969 with Tuf-Tots, ultra-small cars and trucks in the idiom of Jack Odell's earliest Matchbox concept, with tiny packaging to match and aimed at the younger end of the market. The scale was 1:85 for the cars including a neat Dodge Dart and a Citroën DS convertible, and 1:118 for the trucks. At 1s 3d, they were also very cheap and for a while they sold strongly, too. Their old-fashioned manufacturer was slowly ailing (it was bankrupt by 1983) and either couldn't or wouldn't keep up with its bigger competitors.

By the early 1950s, tinplate toy cars were dying out; even Tri-ang's well-respected Minic range was slowly moving over to plastic bodies, with decidedly mixed results in terms of quality. Yet, in 1952, a novel range appeared in shops under the Scalex banner. They came from a new name in the toy industry, Minimodels of Havant, Hampshire, and it was immediately obvious that these 1:32-scale sports cars including an MG TF, Jaguar XK120 and Aston Martin DB2, had lines that were painstakingly designed to represent the real thing. However, they had an added aspect. Their toolmaker designer, Fred Francis, wanted to up the game for 'play value'. Instead of a conventional clockwork motor, with a daft-looking outsized key that could so easily be lost, these had a push-down-and-pull-back mechanism to wind the cars up, rattling off a few yards when they were released.

Popular as the Scalex cars were (7,000 a week were being made in the mid-1950s), Francis was fixated on the idea of cars that could move independently without the direct assistance of a human hand. After seeing an electric motor racing set by Victory Industries, he adapted his own Scalex cars with a gimbal to power them around a continuous slot on a rubber track. Once perfected, he had created Scalex-electric, or Scalextric as it was christened at its 1957 launch. Battery power changed to mains electricity via a transformer, gimbals gave way to a securing peg and trailing wire pick-up brushes, and handheld variable speed controllers replaced on/off knobs. Scalextric would be wildly successful and made a natural progression for die-cast car fanatics when they wanted something more challenging and less childish into which to channel their enthusiasm.

Tri-ang made Francis an offer he couldn't refuse in 1958 and added Scalextric to its massive portfolio of playthings. Just one year later, Lines Brothers' Tri-ang also made a carefully planned full-frontal attack on Meccano's Dinky Toys with its Spot On project. These were die-cast cars and commercial vehicles pure and simple, manufactured in a plant in Northern Ireland. The Lines Brothers (Ireland) Ltd factory, the Pedigree Works in Castlereagh Road, had been opened in 1946 to make prams and teddy bears, and then it was extended in the late 1950s to accommodate the ambitious new die-casting venture. Lines made a concerted effort to include both commonplace and esoteric British cars in the line-up – from the Ford Zephyr and Austin A40 at one end of the spectrum to the Bristol 406 and Jaguar XKSS at the other – and the finish and paintwork were to a very high standard with pleasing highlights such as number plate decals and very carefully designed interiors. The lorries in the range were nothing short of magnificent in their up-close faithfulness to the vehicles that inspired them.

The key tenet to Spot On, however, was the strict adherence to a single scale, in this case the unusual one of 1:42. Everything in the catalogue, from the smallest bubble car to the biggest bus, looked absolutely pukka when placed together in any play situation, along with 1:42-scale accessories including road signs and even pavement sets.

It's probable all the research and development was handled at Lines Brothers' headquarters in Merton, south-west London – a long way from Belfast, but then Corgi Toys were devised in Northamptonshire yet manufactured many miles away in South Wales. Presumably the Merton factory premises was full to capacity, so the Belfast outpost was chosen instead. At Castlereagh Road, the daily production graft divided along similar lines to that found at the die-cast 'establishment' in Swansea, Hackney and Liverpool: the heavy, sweaty and hazardous handling of molten metal was a male domain, while the fitting together of the castings and components on the assembly line was women's work, seated, with a dextrous relentlessness leavened very slightly by the ability to chat over the clatter of the factory apparatus. Contemporary accounts of the work involved in Spot On are impossible to come by but reminiscences among former employees on Belfast-centred online forums paint a picture of personal friendships and family links flourishing despite a hostile attitude from managers, who were interested solely in the amount of work – piece work, indeed – that could be wrought from each person. There seems to have been little that was enlightened about working conditions at 'The Pram', and the finely detailed miniature automobiles that were packed into their 'Spot On Models by Tri-Ang' boxes, complete with their livery featuring graph paper and dividers to suggest a pinpoint obsession with accuracy, could have been just about any sort of widgets as far as those ladies were concerned. They worked there because they needed to.

The range grew quickly and the cars in particular offered a level of realism unseen previously, extending to such tiny items as roof racks, separate grilles and bumpers and even wing mirrors. These fine details were made from plastic that was given a faux chrome plating and were very delicate, meaning most of the later Spot

Ons didn't stay looking spruce for long once they'd been played with by careless boys. Tri-ang overestimated the demands of the market with Spot On, although for ageing children fixated on modelling perfection they had great appeal. Sales were never enormous and in today's collector market Spot Ons are sought very keenly, and can be worth an absolute fortune in the rare cases when they're totally undamaged and in their original boxes. However, before the company took the inevitable decision that their own range just couldn't cut it against the bigger brands, Lines Brothers had bought Meccano and, with it, the venerable institution of Dinky Toys. Spot On would linger just a couple more years before being wound up quietly in 1967. A few of the moulds were shipped to New Zealand, where more were produced, and the Castlereagh Road plant was eventually sold and turned over to manufacturing something completely different: Goblin Teasmade tea-making machines.

Away from the UK, many developed countries came up with a 'national champion' that rose to tackle, in particular, the invasion of the omnipotent Matchbox. In France, this was Majorette, in Germany it was Siku, in Spain Mira, in Japan Tomica. At the larger scale typified by Dinky, notable European rivals were Denmark's Tekno, Germany's Gama and Märklin, Spain's Auto Pilen, France's Solido and Norev, and Italy's Mebetoys, Politoys and Mercury. And there were plenty more, although surprisingly there was almost no direct competition from the USA.

There is one more name that must have a mention here and that is Britains Ltd. This firm, based in Walthamstow, east London, had a heritage in making toy soldiers that stretched right back to 1893 when William Britain Jnr established the company.

Over the years, it has issued all manner of lead and later die-cast vehicles to complement its military, farm and zoo figures, as well as offering several vehicles in its Lilliput series designed specifically as OO-gauge railway accessories. The animals may have seemed to be machine-made and the castings were produced by industrial methods, but every single one was hand painted.

In contrast to Meccano and Mettoy, Britains employed both production-line workers and outworkers based at home. Like those two companies, these were overwhelmingly women. One of the latter in the early 1950s was Alice Maidment, a widow with two young children. Her daughter, Marian Wright, recalled:

> We lived on Lancaster Road and Britains was just at the end of the street – everyone knew it. My mother was a very proud woman and she wouldn't accept any support when my father died, so she worked in the evenings so she could still look after us during the day. Every week, a man arrived from Britains with huge cases of grey, unpainted lead animals, pots of paint and one example of how they should be finished. Then in the evenings, she would paint them at the kitchen table when we'd gone to bed. I've no idea how she did them all because there were thousands, and she would pick out all the detail, like the udders on a cow. Then they'd be collected and the next lot would arrive. My father would never have let her work, but then she had no choice. It was a tough life.

Britains Ltd army and agricultural vehicles were very popular from the 1960s through to the 1990s and made genuine rivals to similar models from Dinky,

Corgi and Matchbox King Size. The key difference, once again, was consistency. Everything from Britains was to an orderly 1:32 scale. This certainly made for some terrific models, often with clever working parts, but it was rare to find anything from the company that was compatible with models from the big guns.

10

Hot Wheels Barges in and Causes Mayhem

It was a guerrilla move of unprecedented audacity – a brazen tactic that could only have been calculated to blast pepper into the complacent, if gentlemanly, faces of the grandees of the British toy industry. The Americans had finally made their move on the die-cast toy car sector and they wanted the upset to be as big as possible.

Rather than reveal its Hot Wheels range in California, where the company was headquartered, or at the International Toy Fair at New York's famous Toy Center, Mattel chose the Brighton Toy Fair in England in January 1968. This, of course, was the home patch of the world's three most successful die-cast toy manufacturers, whose noses would be put seriously out of joint.

Actually, Mattel's arrival in the UK had already happened the previous year. It had bought a British manufacturer called Rosebud, which sold £1 million-worth of plastic dolls a year from the largest factory of its type in Europe. The Mattel acquisition was part of a wider invasion of US firms in sleepy old Britain that saw Ideal set up a local factory and Tonka steel vehicles appear in British toyshops for the first time. Mattel now qualified as a British manufacturer, which allowed it to exhibit in Brighton under the rules of the organising British Toy & Hobby Association. Local rivals, therefore, were prepared for intensified opposition from Mattel/Rosebud dolls but not for an audacious ambush of Matchbox.

And things only got worse for the home team. At the same time as taking the wraps off the first sixteen Hot Wheels cars, Mattel announced that it was starting a splurge of TV advertising to propel Hot Wheels into the ether of every British home equipped with an aerial. They'd spend a then enormous £240,000 on the campaign and it would run throughout the whole year to spread the impact much wider than the usual pre-Christmas push. By comparison, Lesney allocated just £10,000 to advertising that year.

The change in pace would have a seismic effect on the toy car business. As historian Kenneth Brown says in his book *The British Toy Business*: 'One toy retailer later commented that from then on "it was quite uncanny, however poor the toy or amateurish the advertisement, advertised products always sold much better than the rest of the toys".' And anyone who can remember the toy commercials on ITV on Saturday mornings in the 1970s will certainly recall the raucous crudity of the messages, in stark contrast to the sophisticated thirty-second classics (from Hovis bread to Hamlet cigars) that aired in the evenings for adults. By 1970, Mattel was splashing out more on TV advertising than the next eighteen biggest spenders combined in its sector … and Lesney was among those, paying out £200,000 itself for slots between the cartoons. Worldwide, Mattel was reported to have showered $10 million on the introductory Hot Wheels promotion, factoring in an initial huge loss for the anticipated profits later on. 'They brainwashed the kids,' said Odell many years later. 'Every fifteen minutes on American television it was Hot Wheels, Hot Wheels, Hot Wheels.'

Mattel first got serious about entering the toy-vehicle arena in 1966. Elliot and Ruth Handler founded the firm in 1945 and were now both aged 50, and it did not please them one bit to see how much their grandsons loved playing with imported toy vehicles that, although it's unrecorded, were pretty certain to have hailed from Liverpool, Swansea or Hackney! In typically brash style, the power couple wanted their competing range to offer something the top-selling incumbents didn't and they reckoned that the hot-rod and custom-car culture prevalent in California at the time would kick the ass of the worthy subject matter that Lesney turned out – sensible saloon cars, utilitarian trucks or emergency vehicles. Matchbox was a byword for good quality but some of its 1–75 issues could appear a little drab and the inspiration material could be perceived as earnest and European-biased ... although ever-booming sales didn't suggest the strategy was flawed.

To fashion the range, the Handlers hired a proper car designer, Harry Bradley, who had worked previously for Chrysler and had the distinction of designing the world's first concept truck, the futuristic Dodge Deora pickup. Harry liked to commute to work in Hawthorne, California, in his own, immaculately presented hot rod, and when Elliot Handler saw him getting out of it one morning in the company car park, he declared: 'Man, those are some hot wheels.' And lo, a world-famous brand name was born unto us car-mad kids.

Under Bradley's supervision, most of the early Hot Wheels cars would be either customised versions of familiar production models or else inspired by some of the more extreme hot rods he admired. All of them would suggest speed, power and style and, in something of a coup, the Hot Wheels Chevrolet Corvette Stingray was in the hands of the public before the real thing. Bradley's personal customised Chevrolet C-10 Fleetside pickup appeared in the line-up alongside a Plymouth Barracuda, a Cadillac Eldorado and a Ford Mustang. The hot rods included the Hot Heap, Python and Beatnik Bandit, and for good measure there was also a miniature Dodge Deora.

The cars were painted in 'spectraflame', a finish that relied on a polished die-cast body, coated with clear lacquer and then sprayed with such exotic hues as Magenta or Antifreeze. The effect was as if the car had emerged from a custom paint shop with metallic or metalflake bodywork, and several of the range then had matt black roofs to suggest the then highly fashionable black vinyl. Sometimes the paintwork coverage looked a little thin, certainly next to the deep enamel lustre of a Matchbox car, and the Hot Pink colour was soon dropped after market research found the young male customer base regarded it as just too girly.

Hot Wheels weren't just stand-alone toy cars to push along with your fingers or admire in a row on shelf or table. They were designed to really shift on their own, at scale speed of up to 200mph, all without the need for expensive batteries or, as for Scalextric, mains power via a transformer. Mattel created a range of orange plastic track for the cars to run on. It could be clipped together in any way you wanted, although the intended start was via a clamp fixed to a tabletop or chair back, with the track dipping down steeply towards the floor. Simply releasing the car at the peaked start began its gravity-led descent. It would hurtle down the track and then maintain its high-speed momentum to be able to power through loop-the-loop sections and perform other daredevil stunts that could be built into the layout. In the same way Fred Francis had brought life to model cars with Scalextric, now Hot Wheels suddenly leapfrogged Matchbox in terms of the action and independent, unpredictable excitement that could be had from a simple die-cast toy car. There was an wide range of track sets and accessories in the early years as Mattel urged young buyers to 'Race 'em or stunt 'em'.

The secret was contained in the wheels themselves. They were made from hollow plastic that appeared to have a wide tread – in line with the fat wheels you might have found on a pumped-up, full-sized hot rod. Yet due to their slanted profile, only a tiny sliver of the tread had any contact with the surface of a track section or the floor on which they were raced. Allied to thin wire axles (Matchbox used nail-like rods made of mild steel cable) and durable Delrin plastic bushes between these axles and the wheels, a massive level of friction was eliminated, enabling the cars to go like stink. There was spring suspension all round. And just to add an extra Cal-look flourish, the wheels had a red ring around their sidewalls, leading to early issues being nicknamed Redlines.

The cars were packaged in wide, punchy-looking blister packs to hang from shop carousels (like Mettoy's Huskys), each of the 'original sixteen' or 'sweet sixteen', as aficionados later named them, also coming with its own tin collector button. In the USA, the cars sold for 59 cents and in the UK they cost 2s 6d apiece (Matchbox 1–75 models were 2s 4d). Their success was immediate, total and massive.

Hot Wheels, however, were sourced neither from the UK nor the USA but from Hong Kong. Mattel bought a controlling 70 per cent stake in HKI in 1966, whose die-casting track record up to that time had included the brief and unhappy spell as Dinky Toys sub-contractor to Meccano/Lines Brothers. Now, with new American management techniques and quality control, the HKI plant became Hong Kong's most advanced toy factory, developing, for instance, new hot-stamping techniques to produce the shiny silver wheel centres for Hot Wheels. 'Using low-cost Hong Kong was a key factor in its success,' concluded Sarah Monks in *Toy Town*, her definitive history of the Hong Kong toy industry. 'Moreover, Hot Wheels cars made in Hong Kong could at that time enter the US duty free.'

The UK up to this point had been the hub of the die-casting toy car world, and rather surprisingly DCMT Lone Star proved the nimblest in mounting a riposte. Its Roadmaster Impy series of cars was quickly re-engineered with their own designs of low-friction plastic wheels, thin axles and baseplates that were now screw-fixed rather than sealed by a flattened spigot (the steering function was sacrificed, though). They were renamed the Flyers 18-car series and revealed to the toy trade in December 1968 along with an accompanying racing track system called Flyway. New metallic paint schemes with go-faster stripes and a joint-promotion with Quaker cereals that saw some 150,000 given away in return for on-pack coupons got the Flyers noticed at their public launch in April 1970. Stan Perrin, DCMT's managing director, wrote:

We were all the time trying to establish a niche in the market between Matchbox, Corgi and Dinky. Only once did we get ahead of them, and that was when we introduced Flyers, and this only because we were allowed into the Mattel stand at the New York Toy Fair, saw the first demonstration of their Hot Wheels, and decided this was the way to go. We did fantastic business in Japan and the USA.

In 1971, Lone Star was granted exclusive rights to produce a Flyers model of the new Vauxhall Firenza to help launch the rival to the Ford Capri; the company did a super job on it and, because the Firenza was distributed only via the Vauxhall sales network, today it's very rare and avidly sought after by collectors. After that, though, the range suffered a stagnation that saw the early popularity quickly evaporate. The price, at 3s 6d per car, was ultimately uncompetitive anyway at a third more than key rivals.

So much for the also-ran. The effect of the Hot Wheels launch was little short of devastating for long-time market juggernaut Lesney. The company was caught utterly unawares, presumably as Mattel had intended, and was left reeling as sales of its core 1–75 series fell off a cliff. A Kmart order for 50m Hot Wheels cars was just one hammer blow. It was hero to near-zero overnight and in 1969 the real pain showed through. Annual sales in the US, which accounted for 40 per cent of Lesney's output, collapsed from $28 million to $6 million, and half-year profits plunged 29 per cent.

The value of the company's shares plummeted from 95s each to 7s 3d by the end of 1970. In the process, this wiped some £33 million from the value of the family interests of Lesney founders Leslie Smith and Jack Odell, reducing them to a mere £2.5 million. At this point, Mattel had launched another twenty-four Hot Wheels cars, including some non-American models such as a Rolls-Royce Silver Shadow and Maserati Mistral to reflect the brand's global aspirations. HKI's factory at Smithfield in Hong Kong reached a daily peak of 368,348 cars in early 1970, and it could only meet the relentlessly rising demand by outsourcing. Hot Wheels was a phenomenon.

What could Matchbox do? It hadn't exactly been caught napping, but it had certainly become used to its dominant position in the world market for small die-cast toy vehicles. Now it had to rescue itself from financial annihilation and hit the innovation trail simultaneously with renewed energy. That would be a tall order and for Smith there was the added personal strife in 1969 of the death of his wife Nancy.

Lesney's first drastic action was to close four of its fourteen factories, slam the brakes on output and make substantial numbers of the 6,000 employees redundant. Even then, though, the partners would not countenance transferring die-cast manufacture to the Far East to cash in on the low labour rates in factories that, often, were little more than sweatshops. Indeed, in 1969, and despite the prevailing industrial panic, Lesney's newest factory was in Essex, a former Ekco electronics plant it had bought a year earlier. 'News flash!' was the headline for a recruitment advert in a local newspaper. 'The manufacturers of the world-famous "Matchbox" toys are coming to Rochford! Ladies: do you want clean work, pleasant conditions, and good rates of pay? Well, this is your chance. Join us (from April 14th 1969 onwards) and start to change your standards of working and living.'

It was hardly the approach of a rapacious outsourcer and the plant would eventually employ 2,000 people, with the same free bus transport the women of Hackney enjoyed, in this case bringing workers in from local Essex towns including Southend, Shoeburyness, Rayleigh and Canvey Island.

Eighteen months after Hot Wheels cars had hit the shops with such impact, Lesney's response was finally ready to go in August 1970: Superfast was its trendy new high-speed wheel. The company's designers were greatly aided in their work by the fact that the Hot Wheels wheel/axle design had not been patented, so they could achieve the same, low-friction, fast-running result using very thin wire axles (0.6mm diameter, in place of the 1.6mm diameter of the old steel pin axles). According to former Lesney employee Alan Anderson, the Superfast design was created by his foreman Graham Smith at the Carpenters Road, Hackney Wick, plant. Every model in the 1–75 and (from 1971) King Size series that could be modified was adapted to include them; the modifications were often clear to see in the hasty, messy adaptations of the baseplates, and the sometimes brutal reshaping of wheel arches to accommodate wider wheels. Jack Odell did not approve, and thought the Hot Wheels craze would pass and there was no need to corrupt the product range in an overreaction, but he was overruled. The extensive retooling of its ranges subsequently cost Lesney a king's ransom, made possible by an emergency £1.5 million bank loan, but as Leslie Smith later admitted: 'We were on the ropes because until then we had had great cashflow and now suddenly it started to disappear.' All new issues would have the wheels and the ranges were renamed Matchbox Superfast and Matchbox Super Kings. Many of the small Superfast cars received bright new paint colours, including gleaming metallics, and the traditional Matchbox quality was evident over the cheaper and lighter Hot Wheels counterparts. The individual cardboard boxes remained but a new blister pack was also created, while both got a total artistic makeover with illustrations emphasising speed and power, rather than earnest realism. In a corresponding change, in December 1970, the Road Dragster model appeared at No. 19 in the seventy-five-strong Superfast range. It was the first generic car design created entirely in-house at Lesney, a fundamentally ridiculous-looking car with a gigantic, chromed plastic engine sticking out of its frontage. It was soon followed by the similarly 'made up' Draguar and radically modified versions of the existing Mercury Cougar and Ford Mustang, renamed as the Rat Rod and Wildcat dragsters.

Jack Odell, always passionate about accuracy, took a deep breath before chasing the Hot Wheels fantasy car ethos, but his designers dutifully followed the new credo. The Superfast and Super Kings (and associated Speed Kings) ranges were soon peppered with models of actual custom cars and totally fictitious vehicles, resulting in some road rollers, tractors and even hovercraft all exuding a ludicrous customised image. Times had changed. And nor did Corgi or Dinky escape the fury of the Hot Wheels onslaught.

The biggest problem Mettoy faced in 1969 was a disastrous fire at its Queensway, Swansea, premises, an inferno that destroyed its warehouses on 10 March and sent a year's stock of Corgi Toys up in smoke. Many thought the Mettoy company couldn't survive this bitter blow, but in the autumn of 1969 it was firmly back on the innovation trail as Corgi unveiled its Ferrari Dino 206 Sport and Chevrolet Astro 1 with Whizzwheels, hastily retooled cars originally intended for the Take Off wheel treatment but now given low-friction axles. The rolling resistance of these was still impeded by separate rubber tyres, though, and in 1970 Whizzwheels on the new Ford Capri,

Lamborghini Miura, Porsche 911, Iso Grifo and Probe 15 were all hollow plastic with a narrow ridge of surface contact. Most other cars were updated to incorporate Whizzwheels too, but Corgi Track Sets for these large and heavy models didn't work properly and were withdrawn very quickly.

The devastating blaze at Swansea had a profound and unintended effect on Mettoy's international ambitions for Corgi Toys. Sales had ballooned throughout the late 1960s in the crucial US market, bringing welcome prosperity to the concessionaire in the territory, Reeves International. As the 1960s drew to a close Corgi Toys were accounting for 60–65 per cent of the importer's turnover, and Mettoy's management in Britain started to eye this with undisguised envy. Anthony Fleischmann, son of Reeves' proprietor Werner Fleischmann, said:

> Mettoy decided it wanted to acquire Reeves, but my father didn't want to sell his business. It was a bit of a shove and a push, and I don't think it was all that congenial but, in any case, I think my father really had no choice. He had to agree. It's always the same as a distributor: you can't win. If you do poorly the manufacturer will look for someone else, and if you do really well they'll want to take it from you and do it themselves. You own nothing, contracts or no, frankly. That's what it was like with Corgi [in 1969]. With his lawyer he flew to London and they all drew up documents for the closing of the deal.

By extraordinary coincidence, Werner Fleischmann was asleep in his hotel room in London on the night the Corgi warehouse was reduced to a smouldering wreck. The takeover agreement, which had already been signed, was hastily rescinded by mutual consent, and to Fleischmann's delight. 'So the deal fell through and my father came home and continued on as a distributor because they couldn't finance the deal. Things went on as before, and my father continued to build the business.'

Mettoy's exclusive deal with Woolworth's for the Husky series ended in 1968 and this, allied to the Hot Wheels competition, was the spur to rename them Corgi Juniors and convert them to Whizzwheels, too. In 1969, Corgi Juniors appeared in outlets all over the country – Woolworth's included, it should be noted. Yet Mettoy elected to take on Hot Wheels directly with its all-British Corgi Rockets. The first seven were transformations of Corgi Juniors cars, although others would be brand-new castings and paintwork was in mirror-like polychromatic finish. They all had a separate plastic chassis that could be removed with a so-called Golden Key and little tubes of oil so the low-friction axles could be lubricated rather pointlessly. Just like Hot Wheels, there was a huge range of track and accessories so entertaining stunt tracks could be built.

The design work was once again down to John Marshall. In trying to out-do Hot Wheels with something different, he first tried coating the wheels of prototypes with Teflon to coax out the extra speed needed. He recalled: 'We finished up using axles made like highly polished needles, and then two little ribs in the wheels got rid of almost all the friction.'

John was on hand in both London and New York to demonstrate the Corgi Rockets system to buyers from Woolworths, and to Sears Roebuck in Chicago, because he was also the creator of many of the track components – including the starting gate and winning post, the table clamp and loop-the-loop, while the 'Super Booster' that fired the cars on their high-speed starts was the work of his sample-room colleague Tom Chapman. Corgi Rockets seemed to be on an immediate roll, scooping the Toy Trader Toy of the Year award (Boys) in 1970 (the award for Girls went to the Sindy doll) and running a dedicated TV ad campaign. And yet suddenly, at the end of 1971, the whole Rockets enterprise was scrapped abruptly. Why that happened I have never been able to find out, although several people have suggested the threat of a lawsuit from Mattel over track-design patents was the likely reason. One legacy of this sudden end is that certain Corgi Rockets items have become very sought after; the No. 978 James Bond Gift Set, containing bespoke editions of Ford Escort Mexico, Mercury Cougar, Ford Capri and Mercedes 280SL, was sold in 2015 by Vectis for an eye-watering £2,640.

Strangely, the Hot Wheels fad came to a halt just as rapidly as it had sprung up. Mattel had kitted out a seven-storey Hong Kong building in Quarry Bay to turn out Hot Wheels cars at the staggering rate of 16 million a month, with 1,200 perspiring workers struggling to keep pace. In the event, though, this output was attained only for the first three months after its opening in late 1970, and rapidly dwindled to a mere 1 million a month. It was closed in 1973 with 1,000 staff sacked. There were two reasons. First, public demand for Hot Wheels petered out as the car-track craze passed quickly in the far more fickle toy market of the 1970s; and second, prices for raw materials such as zinc and oil-based plastics quadrupled in the build-up to the world fuel crisis in 1973. What's more, any attempt to charge more than 99 cents for a car in the US saw sales judder to a halt.

Matchbox and Corgi survived – just – but Dinky Toys temporarily bit the dust. Lines Brothers had been counting on a £5 million investment from financiers Gallagher in the USA as it struggled to withstand the levels of foreign competition it was now subjected to at home and abroad, and sagged under a colossal mountain of debt. When that plan fell through, the company – with Walter Lines' son Moray as the helpless chairman at the time – rapidly went bankrupt in 1971, and its Tri-ang empire lay in tatters, no doubt to the utter dismay of old Walter Lines, who died the following year at his Surrey mansion.

Dinky Toys, as part of the desperately uncompetitive and lumbering Meccano, were shown up by Hot Wheels as old-fashioned, heavy and humdrum playthings – toys that were expensive to buy and seemed to be from a bygone era when life was slower and the world more ordered. Dinky responded to Hot Wheels with its Speedwheels. Making their debut in 1969 on the Pontiac Parisienne, they initially had low-friction axles supporting cast wheels and rubber tyres. They functioned impressively in the models selected for conversion, although a cheaper, all-plastic Speedwheel was soon standardised. A high-speed track system called Ziptrack was announced but never put on sale, most likely because producing track that could support the considerably heavier (than Hot Wheels) Dinky Toys cars would have been much too expensive and complicated. The brand also dipped its toe into the burgeoning drag-racing scene from which Hot Wheels drew so much inspiration. Dinky's dragster came with free-spinning Speedwheels and a sprung launcher unit to shoot it off towards the horizon, or the skirting board; an unnamed New Projects Manager at Binns

Road, reported in *Meccano Magazine*, came up with the 'Inch Pincher' name lettered on the sticker covering its long, pointed nosecone. Otherwise, Dinky Toys tried to stave off oblivion by axing some of its finest cars – such as the Rolls-Royce Silver Shadow and Aston Martin DB6 – that were crammed with opening parts and hand-applied detail that led to their high prices. The Shadow, for example, required forty-four separate parts. Although Meccano had been rescued from receivership by Airfix, the veteran British pioneer of die-cast toy cars was already living on borrowed time.

11

Struggle of the 1970s: Fighting Back against Star Wars and Teenage Indifference

In 1976, much less than a year after the Triumph TR7 was itself announced, Dinky Toys added a model of it to its catalogue. It was an excellent casting at about 1:42 scale of the controversial, wedge-shaped British sports car. The doors opened and it had a model engineering first – soft plastic bumpers that were 'sprung' to resist knocks, just like the actual car's, incorporated very neatly. What's more, there would shortly be a Rally Car version in authentic livery and later an iteration dressed up as Purdey's car from the popular TV action series *The New Avengers*.

Here was Dinky at its traditional best, right on the button with a pleasing model of the very latest British motoring icon. I was 11 at the time, potty about cars both full and pint-sized, and I can recall vividly getting my hands on one after a frantic burst of extra household chores for a pocket-money bonus.

I could have no idea that within three years the Dinky Toys era would come to an ignominious end. Even its final consumer catalogue in 1979 seemed to boast more new issues than one had done for years, including a good-looking Ford Granada and a Jaguar XJ5.3C for Steed, Purdey's *New Avengers* partner. In the event, both the prototypes so temptingly arrayed to foster desire would never make the toyshop counter.

The purchase of Meccano Ltd by model kits giant Airfix in 1971 had seemed to pluck Dinky Toys from the jaws of death. Lines Brothers, its previous custodian, had failed to make a go of its famous die-cast brand but Airfix was dedicated to the satisfaction of good modelling and delight in miniaturised detail. The first interesting move it made was to launch Dinky Kits. A wide range of cars, commercial vehicles and aircraft were niftily adapted for home assembly by using screw fixings rather than rivets, and they were packaged temptingly with paint and sticker sets. Bearing in mind its Meccano parentage, it was surprising no one at Binns Road had thought of this before, and for older enthusiasts they were rather more experiential than merely a factory-fresh new toy car in a box.

In the mainstream range, though, a slew of rationalisation took place. Many elaborate cars were axed, while others were simplified. Often they lost their opening doors or spring suspension and gained plastic Speedwheels not actually for added speed but to usurp separate cast wheels and tyres cheaply. Most of the new issues tended to be big, heavy vehicles, either for road-making or military duties, and there were further 1:32-scale fighting vehicles that could accompany the plastic soldiers made by Britains and Crescent on manoeuvres.

Meccano's Paris factory had closed in 1970, with all manufacture moved to a new plant in Calais. Then Lines Brothers went bust and all French operations were sold off by the receivers. The new US-based owners of the French assets decided die-cast toy vehicles were of no interest to them and most of the tooling was sold to Spanish die-casting company Auto Pilen in 1972. This company then continued making and packaging French-designed Dinky Toys for the French market. However, just before these international shenanigans began, a British 'raid' on the French range had taken place, and five model cars suddenly reappeared on the 1971 production lines in Liverpool. They were the Matra 630, Ferrari 312P, Opel Commodore coupé, Citroën Dyane and Fiat-Abarth 2000 Pininfarina, a distinctly European line-up, and bagging them meant there were no new traditional road-car British issues in 1972 or '73. All five were to the 1:43 scale that French Dinky Toys had stuck to faithfully since the 1950 arrival of its Peugeot 203. Out of step with Dinky's adoptive 1:42 scale, they were a cheap shortcut at a time when the whole business was on the ropes.

Lines Brothers' US distribution arm had vanished with the rest of the collapsed company. Cowell Management, a Californian company, then took on US Meccano distribution, which included Dinky Toys. From a modest start it did such a solid job that it was taken over by AVA Industries of Waco, Texas in 1973, who would then import and market Dinky's vehicles for the rest of their existence. This key distributor never seemed to exert much influence over new issues, with only a small handful of upcoming Dinky releases reflecting the US automotive scene; a fire truck from the TV series *Emergency* and a Plymouth Gran Fury as a police car or New York taxi was pretty much all that emerged.

Italian die-cast manufacturers such as Burago and Polistil had found a ready market for highly detailed 1:25-scale cars that were fun to play with and also reproduced faithfully enough to make great display items for older petrolheads; now Dinky tried to get in on this trend with its own 1:25 range in 1973. Such as it was. The Ford Capri MkI could be had as a standard coupé, a rally car or as a police patrol car – the detailed livery for this one derived from the Lancashire Constabulary, whose patch Meccano's factory was within – yet for some reason they didn't sell at all well and no further issues were added.

That didn't deter Dinky designers from thinking bigger for other cars. As Corgi had upped the size of its mainstream cars to 1:36, so Dinky now followed suit and the Volvo 265DL, Princess 2200, Plymouth Gran Fury and Hesketh 308E Formula 1 car of the late 1970s were all much more chunky than previous approximately 1:42-scale cars.

At the start of Airfix's proprietorship things looked hopeful. Meccano made a profit of £335,000 in 1971–72 and in 1973 Dinky was still the fifth best-selling range at British toy retailers and the Liverpool company's most important product line. However, problems were fermenting in the Binns Road factory. Incompetent management led to deteriorating industrial relations and trade unions took this opportunity to instil a previously unseen militancy in the workforce. Ironically, it was an exact repeat of what was happening in the car industry itself, exemplified by the misfiring shambles of British Leyland. Heels were dug in on working practices. For example, although new machinery was delivered to the plant to remove flashing from die-cast vehicle bodies automatically prior to painting, the female workers refused to use them, insisting the work continued to be done by hand. Managers were frequently long-serving and poorly trained, and often just as resistant to changes around the works as the assembly operatives.

Once the increased price of raw materials was factored in, costs started to spiral. Internal figures show the retail price of a Dinky Jaguar E-type rose from 99p in June 1974 to £1.40 by July 1975, while an AEC Esso petrol tanker went from £2.25 to £3.25 over the same period. By comparison, over the same twelve months, Corgi's E-type rose from 75p to 90p and its Mack Esso petrol tanker rose from £2.15 to £2.75. Not surprisingly, customers noticed these differences and began to steer clear of Dinky.

Meccano's old solution of manufacturing abroad was tried again. The Rover 3500 SD1 was an exciting new British car in 1976, but it took until 1978 for Dinky to offer its own 1:36-scale version, sourced from Hong Kong, and even then it was a shoddy casting with poor paintwork. It was the same for a rather horrible Hong Kong-made Jaguar XJ5.3C; a small number were distributed, not even in a box but in bubble-wrap bags and the *New Avengers* edition of the same car hit so many quality snags that it never reached the market at all.

Meccano also attempted to tackle Tonka head on with its Mogul range of heavy steel toys for very young children. In the 1960s and early '70s, the company only had four designers assigned to Dinky Toys research and development, and so it was an indication of how little emphasis was placed on new products that the Mogul design had to be handled by Ogle, a Letchworth-based consultancy that had shaped everything from the Chopper bike to the Reliant Scimitar car. Mogul, moreover, turned out to be merely a mediocre Tonka copy and it was more expensive than its main rival, so it was doomed from the start.

Ogle was later employed once again to design a range of basic Dinky Toy lorries, much to the delight of its managing director and design chief Tom Karen, who had always had a passion for creating toys. Indeed, at the very same time as overseeing the design of this Dinky 'Convoy' series – similar to the 22 Series of the 1930s in employing a single chassis casting with a variety of different rear bodies – Ogle Design was designing the T45 Roadtrain range of real-life lorries for Leyland Trucks ... and Ogle's cab design for the Leyland Roadrunner light truck of 1984 bore a striking resemblance to the Dinky Convoy launched seven years earlier. Yet another designer of actual vehicles, Tony Stevens, was brought in to fashion a vintage-style Dinky Toys taxi cab after the brand tied up a promotional deal with McVities for its new Taxi chocolate wafer biscuit. Like the Convoy trucks, this Happy Cab has never been shown much love by serious collectors, but their unusual backstories at least make them distinctive pieces.

Corgi's 1970s fortunes were the inverse to Dinky's malaise. In 1971, Mettoy made its first ever trading loss, at £441,000, as material costs rose and its sales slumped while being battered by Hot Wheels. The directors were quick to act, though, shutting promptly its showpiece Northampton facility that made Corgi Juniors (they were shunted off to Swansea) and issuing redundancy notices to the 900 people who worked in this multi-storey former shoe factory in Stanley Road. Company founder Philipp Ullmann died aged 88, while his son Henry and Corgi overlord Howard Fairbairn relinquished their executive powers and new managing director Bernard Hanson was appointed from outside the firm.

The company's fortunes recovered quickly and turnover leapt from £9.3 million in 1972 to £19.9 million by 1976. In October 1973, deputy chairman Arthur Katz told the *Daily Express*: 'We are working flat out to meet the orders. Our only problem is that we are finding it difficult to get the materials to make the toys and the staff to work the hours.'

Katz was honoured again by the Queen in 1974, this time with a CBE. Sales that year rose from 8.2 million to 9.3 million units, the company made almost £1 million in profits and it even had to rent extra factory space in Neath, South Wales, to satisfy Corgi Junior demand, as well as building a new die-casting foundry at Fforestfach. Such was the yo-yo nature of toy-making. In 1974, sales and profits were slightly up again and in 1975 and 1976 profits rose despite lower production output.

In tandem with its quick-witted corporate reactions, this was a product-led recovery. After fifteen years, Marcel van Cleemput was still the creative force behind everything new carrying the Corgi name. Although the general public knew nothing of him, he was already legendary in the toy industry and his spirit was embodied in part of the 1971 promotional campaign with a set of fictional cartoon characters called the Corgi Technocrats: bespectacled H.W., the backroom boffin; Whizz, the fresh-faced kid representing the end user; Penny, looking out for the girls' interests in toy cars (whatever they were ...); and Zak. 'I'm the troubleshooter to the team', went the blurb. 'Whizz and I fly all over the world looking for new ideas, and if there's trouble I sort it out!' These were the characters from *Scooby Doo* and *Mission: Impossible* combined, to reassure young customers that Corgi was giving them exactly what they craved. Van Cleemput would identify and develop a theme so that Corgi Toys could capitalise on it. The informal group of 'dragsters' that joined the Corgi range in 1971 was one of his ideas, inspired by the British craze for these dramatic, acceleration-biased slingshots that had its spiritual home at the Santa Pod Raceway in Bedfordshire, a short distance from the Mettoy Northampton design nerve centre. One of the cars, a Jaguar-powered Austin Seven called Wild Honey, was actually a car built by a couple of Northampton likely lads that van Cleemput chanced upon, liked, and immortalised in table-top form ...

The most significant shift in the ethos of the range brought a move to a much bigger, uniform, 1:36 scale. The change began in 1973 with the Ferrari 365 GTB/4 Daytona and this was followed by the Jaguar XJ12C, Citroën Dyane and Mercedes-Benz 240D. When I saved up to buy my own example of this Merc, I discovered to my dismay that the boot lid was loose because of a manufacturing defect and I decided to write to the factory to complain. I was amazed when, by return of post, they sent me a free replacement. It made me a Corgi devotee, of course, because I couldn't help but feel the Corgi Technocrats really did care about their customers. The Ferrari was especially interesting because a real-life car, driven by JCB tycoon Anthony Bamford, carried Corgi sponsorship into endurance racing in the 1973 Le Mans 24-Hours race. There was also a superb grid of 1:36-scale Formula 1 cars, of which the favourite for many will always remain the Lotus Type 72 John Player Special. This black and gold beauty sold 2.2 million copies. Van Cleemput had to negotiate hard to avoid paying royalties for this and other F1 cars, particularly the Elf Tyrrell, but ultimately all teams and sponsors recognised the promotional value of the models mattered more than any small financial gains.

Corgi, like its rivals, had to face one stark fact that went with the 1970s territory of childhood: children were simply not into toy cars in anything like the way they used to be. There were several influencing factors, such as the emergence of the teenager as a rebellious new force, more TV channels and a rise in programming for children, battery or electrically operated toys, and even 'dolls for boys' in the poseable, macho form of Action Man. The result was, as van Cleemput said:

> During the first few years of Corgi, children and youngsters were buying or having die-cast models bought for them up to the age of 14, 15 and even 16. By 1971, the upper age limit had dropped to 11-year-olds. To make matters worse, by the end of the 1970s, this would be seen to drop down to eight-year-olds.

Still, military hardware remained a big hit with little boys of all ages in the 1970s and Corgi made sure it had tanks and aircraft in its line-up by taking on ranges originated in Hong Kong by Lintoys; to give them a competitive edge with Dinky Toys counterparts, Corgi engineer John Marshall gave the gun barrels on the military vehicles a fantastic recoil action to accompany their shell-firing ability, for which he was named yet again on the patent application. Unlike many other Far East imports, these were of decent quality and they sold well.

More ingenious model engineering could be found on Corgi's 1974 Chubb Pathfinder Airport Crash

Tender, which featured a working water cannon and siren. It cost £90,000 to develop and merited its own TV advertising campaign, which led to an instant sell-out. Pent-up demand was so high that another exercise in cutting costs turned sour. The internal electronics depended on a circuit board sourced from Hong Kong and, with all Pathfinder stocks sold, the six-week wait for more, and the high reject rate, this made it frustratingly hard to capitalise on the euphoria.

Two 1:18-scale Formula 1 cars were created, the JPS Lotus and a Marlboro McLaren, which looked superb and were supplied with no qualms whatsoever about the effect of cigarette advertising on children. And there were myriad other cars and trucks issued each year, individually and in attractive gift sets, together with a brief return to Marks & Spencer with a range of repackaged, St Michael-branded Corgi vehicles.

Gene McKeown worked for importer Reeves International, joining the company in 1973 after Werner Fleischmann encountered him working as the assistant toy buyer in the gigantic Chicago department store John Wanamaker:

> Mr Fleischmann was just a phenomenal salesman – he had a sort of way into outlets that other people couldn't get. He also worked very closely with Peter Katz at Mettoy, especially on logos, advertising and so on, so they both understood the market better than most. They shared a lot. They'd both had a tough time in the War and then they had an opportunity to reinvent themselves because they were entrepreneurs. They took risks but they also enjoyed every day of their lives, they liked a drink and a party and it wasn't always about money. I mean,

they wanted to make a profit, yes, but they enjoyed creating things. One year, Reeves flew the entire salesforce to London to show them what's involved in making Corgi Toy cars, and then chartered a British Rail coach with a bar in it filled with liquor to take them to Swansea on the train ...

I was involved primarily in operations and at the time Reeves had a showroom and offices on 5th Avenue in New York City, right downtown – everybody who was anybody in the toy industry had a showroom round there – with five people including our vice president of marketing Gary Fisher. He primarily handled our European companies like Mettoy along with the owner Mr Fleischmann. I went to the Corgi factory in Swansea in 1976 for about a week, seeing how they operated. It was amazing. There were thousands of people working there. Then I went back to London to the Britains factory in Walthamstow, and I remember thinking that, Oh, my God, this is so old and archaic compared to Mettoy. The Britains factory even referred to itself as a cottage industry, and that was absolutely right!

Tony Fleischmann, Werner's son, recalled his own visit to the Corgi factory in Swansea in 1977:

I spent a semester at Oxford University and while I was there I took a drive down to Swansea with a Reeves colleague, to spend the day touring the factory. We slept above a little pub nearby somewhere. Barry Frears of Mettoy gave me the tour, probably the nicest man on the planet. I thought it was fascinating. I had never really seen a factory like that. I was scratching my head and thinking: why is this place located at the end of the earth? It was raining constantly, and I could hardly understand what anybody was saying! Today I liken it to the restaurant business: you have a wonderful experience with the food, the presentation and everything, but rarely would you want to go into the kitchen and see what's going on there.

Fleischmann's visit was at a hugely rewarding point in the Anglo-American partnership. Reeves' volume had grown 266 per cent over the previous ten years powered by what it called 'carriage trade' toy brands such as Corgi. In a 1978 report in trade magazine *Toy & Hobby World*, Werner Fleischmann put that down to increased prosperity among US consumers, with mass-market retailers taking advantage by stocking more prestigious brands such as Corgi and Britains. They benefited too from some widely reported safety scandals afflicting cheap toys, and the disdain of intelligent, middle-class parents for the trashy nature of many toys advertised on TV. Corgi Toys were seen to be of a superior quality and to encourage brain-feeding play.

The key range that boosted Reeves' turnover were the small-scale Corgi Juniors. In the UK these had long played second fiddle to the all-conquering Matchbox Superfast series, even though there were often interesting car models in its line-up; highlights were the Reliant Scimitar GTE, Jaguar XJ6, Jensen Interceptor, GP Beach Buggy, AMC Pacer, Fiat X1/9, NSU Ro80, Aston Martin DBS and Porsche 917, to take a few personal favourites at random. None of these was ever tackled by Matchbox but could be had as 1:65-scale Corgi Juniors for roughly the same money. Corgi Juniors were now Reeves' passport to be stocked in KMart, Sears, Roebuck and Ben Franklin stores across the US as 'quality-oriented merchandise'. Feedback from Reeves implored Corgi to keep finding new TV tie-ins, and in the 1979 Corgi Junior line-up there would be an explosion of character vehicles linked to TV shows such as *Kojak*, *Starsky & Hutch*, *Wonder Woman* and *Charlie's Angels*, and broadening out the *Batman* franchise to include adversaries like The Joker and The Penguin. Reeves also pressed Corgi to up its game on the packaging front. 'We feel the traditional European approach often does not maximise sales appeal in the US market,' said Reeves' director of marketing Mike Gatto. Indeed, he got so frustrated with Corgi for not developing a collector's case for its Juniors to rival the Matchbox Superfast item that Reeves produced their own exclusively for the USA. 'It is the kind of product that enhances Corgi sales considerably in this country,' Gatto added. He was right: there was nothing like those forty-eight empty slots in the case to encourage buying the latest issues. Some 80 per cent of Reeves' mass-market sales were down to Corgi Juniors alone in 1978–79.

Peter Katz, now Mettoy's managing director, declared 1978 the best ever year for Corgi, with a £3.4 million profit and sales up to 8 million units. Peter had taken over the family mantle from his father Arthur, who retired in 1976, one year after Howard Fairbairn, the original Corgi die-casting expert of twenty years' standing. However, just a year later, the profit collapsed to £723,000 in the face of a strong pound, which hammered exports, while rising interest rates and wage bills bit into costs. As chilly economic winds swirled across the Western world, orders were massively down.

How a Matchbox Car Did the Trick

I'd been particularly objectionable to my long-suffering mother one overcast day more than four decades ago, wittering about some perceived injustice – 'being a bloody nuisance', as my dad used to put it. I was probably not getting my own way and therefore playing up.

Yet, with an astonishing grasp of reverse psychology, my mum could see only a daylong headache ahead unless she could shut me up. Her plan was simple. She stopped the car. She went into a newsagent. She came back to the car. Something blue and yellow was peeping out through her fingers.

It was a stark choice. 'Are you going to behave?' she said over her shoulder to me on the back seat. The downturned corners of my mouth melted away. I nodded bashfully. A small cardboard box was placed in my little hands, still damp from rubbing tear-soaked eyes. It was a miniature Cadillac ambulance, a bit like the one from *Ghostbusters*, with sirens and lights on its roof and, just visible through the rear window, a tiny white plastic patient, prostrate on a stretcher, moulded into the interior.

I can remember what it cost, too. Seventeen New Pence, recently converted from 1s 9d in those freshly minted decimalisation days – this was 1971. It was made by Matchbox.

However crude this hand-held Cadillac was, however unrealistic its Superfast wheels or completely devoid it was of opening, working or exploding features, it did the trick. I played with it for hour upon hour, 'driving' it along the arms of armchairs or across the mountainous terrain of my bed, screeching it around hairpin bends on the floor. Mum got some respite and I had an important new addition to my toy car fleet. And by the way, Mum, sorry for being such a pain ...

The strain of retooling its factories to give Superfast wheels to its small and large vehicle ranges took its toll on Lesney. By late 1970, the company was running at a £700,000 loss and, worse still, it owed the Inland Revenue £2 million from profits made in those late 1960s heydays. 'All of a sudden,' Odell noted, 'we were arguing whether we could afford to buy a new typewriter.' The company's banks were very unhappy about extending credit to a business that, not long ago, was awash with cash, but once Jack Odell had convinced them that they had a proper rescue plan, they reluctantly agreed.

As well as closing factories and reducing headcount by 900, Lesney now intended to diversify, to spread the risk from the die-cast toys through which it was known. To that end, in 1972 it launched Matchbox Kits – plastic construction kits to compete with the very successful range made by Airfix. Odell hired Maurice Landi as the kits' project manager, who came up with twenty aircraft models for the initial range. Meanwhile, to increase its profile and signify its association with desirable fast cars, the Surtees Formula 2 team sported Matchbox sponsorship in 1972. Lesney branched out further in 1973 with a game called Cascade, introduced some plastic pre-school toys, and then in 1975 it launched Fighting Furies – pirate action figures. In 1977 Lesney made a serious play for the attention of the girls' market by acquiring Vogue Dolls from Tonka. And then, in 1978, the company made its biggest ever acquisition by buying America's model-kit maker AMT.

Back in 1973, Jack Odell had decided to retire as joint managing director, leaving Leslie Smith in sole executive charge, although he retained a part-time role as deputy chairman. After having worked like a trojan on his business, he now enjoyed ten holidays a year and almost constant golfing. It turned into a very dicey year for Lesney. In the first few months all production was halted because of an eight-week strike by coal miners, which cut energy supplies to non-essential industries. Management-worker relations at Lesney had always been good thanks to co-operative works councils, but that ended in 1973 when there was a strike in the company's fettling department. And to top it all, the Rochford factory first caught fire, laying waste to millions of plastic toy car parts, and then flooded, putting die-casting machinery out of action. As a result, Lesney suspended Models of Yesteryear manufacture for two years, while the repaired Rochford plant was turned over to the all-new product lines.

Thankfully, for those who loved cars, Lesney refused to abandon its roots. The challenge from Hot Wheels might have abated, but the company's designers kept hard on the innovation trail. With excellent quality, they ventured ever deeper into the realm of fantasy cars, with most new models having a custom car theme and sporting out-sized plastic engines with banks of exhaust pipes. The vivid colour

Matchbox's Guildsman: a Design Apart

Lesney strived to make the vehicles in the Matchbox 1–75 series an interesting and varied selection. Essentially, there would be something for every taste and interest. In the 1970s, as the Hot Wheels threat was beaten back, the cars in the range became an esoteric mixture of real-life subjects and fictitious models from the increasingly imaginative minds of in-house designers. Occasionally, though, there were curious interlopers, such as the No. 40 Guildsman.

Between 1965 and 1970, Vauxhall Motors ran a public design competition called the Craftsman's Guild to tease out new talent. A keen teenage stylist called Phil Gannon entered models three times and eventually secured a third place in 1968 with his 1:12 scale concept for a futuristic sports car. He recalled:

> The written specification included front engine, four-wheel drive, centre roof aerofoil, air con and an infra-red screen in conjunction with infra-red front lighting. I carved my model from a block of Canadian yellow pine and carved moulds likewise to form the Perspex screens, which I made in my mother's oven. Vauxhall supplied the tyres but the alloy wheels were cast and machined by myself. The interior was trimmed in chamois leather and the exhaust pipes were made from copper cable finished with .22 calibre cartridge cases!

Lesney heard about Vauxhall's contest and asked if it could assess all the entries, winners or not, and then pick one to turn into a mass-produced toy. Of all the numerous models that had been entered over the five years, it chose Phil's and the Vauxhall Guildsman was duly issued in 1971 at about 1:64 scale.

'Vauxhall had the publicity rights,' Phil added, 'but two years after the competition I was awarded £25 for the model being produced. Matchbox painted it pink but the original model – which I still have – was gunmetal grey with a black roof.'

palette used – the pinks, oranges and lime greens – started to account for some of the 3,000 gallons of paint the factories consumed every week. I must admit, at the time I hated this kind of thing, but they do all now have real period charm, and they certainly helped Matchbox re-establish itself in the USA as it pummelled Hot Wheels on its home turf. In fact, sales of the new Superfast 1–75 series in particular soared again, both in the UK and the USA.

Rolamatics was a new variation to be found sprinkled among the 1–75 Superfast ranks – cars whose parts moved in various ways as their wheels turned. Clever internal structures gave each one an ingenious feature – from a rotating look-out soldier in the Weasel armoured car to a driver who bounced up and down in the Beach Hopper beach buggy. The mechanisms, however, rarely added to the total number of components in each of the 1–75 vehicles. That was on average thirteen separate metal and plastic parts which, in a typical model car would be four plastic wheels, two steel axles, a die-cast body, a die-cast baseplate, a plastic suspension spring, a one-piece moulded plastic interior unit, a one-piece plastic glazing unit, one or two small die-cast opening panels such as a bonnet or two doors, and perhaps a decal or sticker. Some cars might forgo an opening feature to allow another vehicle in the range to have more parts, such as the AEC 8-Wheel Tipper, which had eighteen. Across the whole, ever-changing range, there were numerous examples of single components fulfilling two roles, such as the AMX Javelin where a black plastic insert provided both the bonnet scoop and the steering wheel. Deeply tinted windows often meant an interior could be left out to allow an extra component to be added elsewhere. It's obvious that, for a Lesney designer, the discipline of staying within strict pricing boundaries that would guarantee profits was the continuous challenge, along with trying to enhance each vehicle with cute details at no cost.

Then there were Streakers, decorative makeovers of existing 1–75 models using a new 'tampo printing' process to place elaborate ink decals on to painted metal, using rubber pads. This certainly brightened up some old castings including the Citroën SM and Bertone Carabo, even if the results were anything but true to life.

Super Kings was the range, quite a considerable one, of larger commercial vehicles – lorries, transporters, construction vehicles, tractors and more – that directly rivalled the main Corgi and Dinky offerings. Working features, interesting cargoes, accessories, articulation and trailers were allied to lustrous finishes, chrome-effect plastic, big free-running Superfast-type wheels and window-fronted packaging with pulse-raising artwork. In parallel was the Speed Kings series built around a few standard road cars and rather more dragsters, motor racing and combined sets of the same, redolent much more of the custom-car culture of California than the humdrum roadscape just outside the Hackney design studios where they were all created. There were usually little touches intended to delight the end user, like the lift-up bodywork on the Gus's Gulper dragster and the interior under the lifting roof of the Camping Cruiser. Realism was deliberately pushed aside for a high-octane fantasy world of outsize exhaust pipes and mag wheels. Or, at least, most of the time; some efforts were made to put real-life showstoppers into junior sweaty palms. Take, for instance, the Mercedes-Benz C111. This was an experimental car that, in the real world, was created by the German manufacturer to conduct further research into the super-refined rotary-engine designs of Dr Felix Wankel, and to delight motor-show visitors who could witness its dramatic gullwing doors opening and closing in slack-jawed admiration. Perfect subject matter, then, for a toy car.

It's one of the few cars that was modelled in period at roughly the same 1:43 scale by Dinky, Corgi and Matchbox, and makes a fascinating example of the different approaches taken. The Dinky Toys model came first in 1969 – good representative proportions and boasting both working gullwing doors and an opening engine cover to reveal the cast representation of the rotary engine mounted in the middle of the car. Letting it down was a totally inauthentic claret body colour and a gold-painted engine, and a length that appeared very slightly stubbier than the real thing. A year later came Corgi's rendition. The profile was more accurate in terms of length, the orange colour was right, and the gullwing doors functioned properly. Yet the nose was a crude, featureless version of the real car's, and the Whizzwheels looked nothing like the intricate alloy wheels of the Mercedes original. There was a small-scale C111 in both a Corgi Junior and Corgi Rockets format, a vague replica where only the engine cover could be opened. Finally, in 1971, Matchbox Speed Kings joined the party and so the car would be in all three ranges concurrently for a year or so. This was the car refracted through the eyes of a toymaker rather than a modelling perfectionist. The stance, proportions, wheels and colours were all considerably adrift from reality – it was pretty much a three-dimensional cartoon. On top of that, the USP of the actual car, those doors, had been eschewed for the Speed King's own gimmick – battery-powered working retractable headlights that could be popped up using a little switch underneath. Only about 100,000 of these were made before production problems with the electrical parts forced Lesney to redesign the car with manually operated headlights without their toy beams. And yet, despite all my criticism here, the Speed Kings Merc offered the most lustrous, vibrant finish and was arguably the more temptingly presented on the toyshop shelf.

Elsewhere, careful examination of these bigger, shinier Matchbox Speed and Super Kings toys also revealed a few in-jokes that were no doubt completely lost on almost everyone outside the factory. For instance, on the cowled nose of the absurd Super Kings Mod Tractor (it came complete with matching trailer similarly emblazoned with garish stickers) was moulded in the fictional make LLEDO ... which was, of course, the backwards-spelt name of the just-retired Matchbox die-casting guru Jack Odell.

This Mod Tractor had the input of Brian Mawdsley, who joined Lesney as a designer in 1976, aged 28. In an interview recorded in 2011 for the Museum of Childhood's British Toy-Making Project by researcher Juliana Vandergrift, he recalled the design department pecking order that he slotted into, with the managers and project engineers above him in salary terms, and the draughtsmen, modelmakers and pattern makers below him:

> It was a really tight unit and we all got on really well. Every Thursday we'd go and play golf in the summer or snooker in the winter, then to the pub, then to have an Indian. I got better at doing drawings that reflected reality. You can cheat with a one-off prototype that works perfectly and looks exactly as it should. In production, everything has to have a tolerance, and we'd work to two thousandths of an inch, to make it produce-able and assemble-able in ten seconds.

Brian was very unusual in also having a spell working for Corgi, and noted the contrasting atmospheres between the two companies.

> Matchbox felt more like an upstart and Corgi felt more an old family business. Where [the] Matchbox [factory] was built in the late 1950s, the Corgi building was 1800s or something. Corgi had three canteens, one for the directors, one for the management and one for the workers, and it was important which canteen you went to. It amused me when Peter Katz answered the phone in his office and said: 'Katz, P.' You got the impression with Jack Odell that he was still, you know, one of the lads.

Battle Kings was a new series of military vehicles in 1975, which was the year that Yesteryears returned (priced at £1.25 each), albeit now in corny-looking woodgrain boxes with the cars painted in resolutely non-authentic colours, Odell no longer being around to put his foot down. Two years later along came the Adventure 2000 range, Super King-sized vehicles that were vaguely space-age in their theme – adopting some of the general Thunderbirds-type combat themes and forms around missile-firing and cresting rough terrain, but without any particular backstory to put them in context. It was yet another 1970s product success story.

The company was thriving once more. Mary Anderson's son, Alan, even decided to follow her path to work for the company – not as a part-time shift worker but as an apprentice:

> I left school in 1975, three years after mother left for another part-time job but with more sociable hours and joined Lesney as an apprentice toolmaker – whether or not having earlier family connections helped I don't know. Essentially a toolmaker makes the machines that make the product, the models, and I worked with some wonderfully skilled blokes who could make something out of nothing. However, earning just £16.36 a week, the lure of British Rail's £27.50 was too much to turn down in 1977 and I changed trades, although still in engineering.

Lesney's flurry of activity seemed to do the trick. The crisis passed and from 1975–77 sales went from £32.5 million to £56.4 million to £88.9 million. Profits returned in 1978, when the company made £5.4 million, its 6,000-strong workforce having just contributed to the company's fifth Queen's Award (by this time it was said to be for industry, rather than export). With new factories once again being added to the company's estate in 1978 it could boast of an astonishing 1.5 million sq. ft of manufacturing space, and all of it in Britain around east and north London and Essex.

Lesney's strong performance in the mid-1970s, however, would again hit the skids. Acquiring the AMT kits business for £4.8 million was expensive and required borrowing at high interest rates. And then strong Sterling, which by 1978–79 was running at an exchange rate of almost $2 to £1, led to offputtingly high prices in export markets. The company anticipated a profit of £9.9 million for 1978, but it turned out to be just £6.8 million. All along the way, toy cars made in places like Hong Kong enjoyed tariff-free access to the crucial US market, while Lesney had to pay 12 per cent duty on everything it exported from England. The odds against it were piling up.

Youthful collectors such as me were, of course, largely unaware of the unfolding industrial turmoil at the time Margaret Thatcher's Conservatives swept to power in May 1979. We were simply lapping up the never-ceasing stream of new Matchbox releases – constantly pondering on what to get next with whatever spare cash we had.

For rather more than a year before this date, the design and product ethos at Matchbox had made a subtle shift back towards realism. Ever since Lesney's prosperous world had been blindsided by Mattel's introduction of Hot Wheels, faithfulness to the subject matter had taken a back seat and an urgency to plug into the overall US West Coast car culture from which Hot Wheels drew its inspiration was an imperative.

The Infamous 'Crested Liptons'

One of the more interesting Models of Yesteryear to take its place in the sixteen-strong range in 1978 was an attractive model of a very unusual van, a 1927 vehicle based on a Talbot chassis. It carried the insignia of Lipton Tea and a 'by royal appointment' crest. This crest needed approval from the Lord Chamberlain's Office, which Lesney assumed would be a mere formality, and cartons of the vans started to be despatched before the letter of agreement was received. When it did arrive, there was a shock: permission was, for some reason, denied. News spread of Lesney's panic that it might be forced to recall the 100,000 examples already with retailers; collectors, sensing an impending rarity, snapped up any they could find. In the end, no recall was forced on the embarrassed company, although the crest decal was immediately changed for one showing a monogram. Funnily enough, the 'crested Liptons' never did become very valuable because survival rates in perfect condition are so high. However, the affair did lead Lesney to focus more on commercial vehicles with different liveries as a way to appeal to Yesteryear collectors in the 1980s. Plentiful contrasting versions of this Talbot, Ford Model T vans and tankers and a Ford Model A van festooned with household brand names and liveries (many not exactly as you might have seen them in the 1920s and '30s, mind) kept avid collectors coming back for more and more, and they tended to out-sell simple Yesteryear cars by up to 40 per cent.

Vehicles tended towards customised or hot-rod themes, or else were entirely fictional creations. Efforts were made to find unusual, full-size, one-off cars to shrink down to small scale such as the Superfast Freeman Intercity Cruiser, the Guildsman (see full story on p. 173) and Siva Spyder. But many more seemed like fun projects for the designers – toiling away in drizzly Hackney rather than sun-kissed California – that made it from imagination to model prototype entirely at the drawing board. You got the distinct feeling that many of them started in a meeting with merely a catchy name – Toe Joe, Stretcha Fetcha, Mini Ha Ha, Fandango, Shovel Nose, Hairy Hustler, Volks Dragon – before the madcap design was produced to fulfil it. That was in the Superfast range; among Speed Kings the thinking was similar and many of the cars – Bazooka, Marauder, Bandolero and Shovel Nose among them – were an affront to any half-serious car enthusiast. Our younger brothers, mind you, probably simply just loved playing with them, dazzled by the intense colours and big, jacked-up back wheels; hours of imagination-bending good fun was guaranteed in every addictive one.

In 1978, all of this started to change. With the choices in both the commercial vehicles of the Super Kings range and the cars and trucks in the huge Superfast arena, Matchbox models started to be built around the 'real thing' again. Pretend vehicles started one by one to be deleted and a new generation of decent models of interesting contemporary vehicles took their place.

The Ford Escort RS2000, Renault 5, Rolls-Royce Silver Shadow II, Porsche 911 Turbo, Ford Transit and Holden pick-ups, and Datsun 260Z tried hard to recapture the lost charm that had made the 1960s issues so irresistible and they largely succeeded. They were plainer issues, lacking the more fantastical features that, because they had tended to be in plated plastic, were soon broken on the previous generations of dragsters and fantasy machines. Materially, too, there was a significant change. Sprung suspension had been a range staple feature, in cars anyway, since the early 1960s, but throughout the 1970s the slimness of the Superfast axles made them easy to bend or kink and indeed, dislodge from their internal mountings so the car would neither roll along nor display well. There was no suspension on most of the new breed from 1978, but then tight internal locations and slightly thicker metal meant they were much harder to damage or distort. By 1981, there were barely any issues in the annual catalogue that didn't have a basis in reality, while retaining a toy-like charm, whereas back in 1975 and '76 the offerings were about 50 per cent fictional.

At the bigger size, a dose of probity came to the Super Kings series (the Speed Kings name was dropped for 1979). Handsome and detailed models of trucks and commercial vehicles started to abound, with Peterbilt and Bedford cabs that were recognisable as just like the ones you might encounter on US and UK roads. Interesting liveries and accessories brought many of these fine models to life. Among cars at roughly 1:36 scale, there was a lovely Jaguar XJ12 available as a police car twinned with two motorcycle outriders or in Civvy Street guise with a Europa caravan, while a Porsche 911 Turbo came with possibly the smallest working detail ever at this size: fully functioning door handles. A companion Volvo 244 estate had intriguing details too, deformable bumpers and a sunroof as a black roof sticker.

However, anyone who received a copy of the new *The Matchbox Annual* in their Christmas stockings in 1979 or 1980 must have wondered what on earth was happening. These books, published by Purnell and presumably intended to spread the Matchbox word, contained plenty on vehicles and machines but inexplicably almost nothing on the toys themselves.

12

Sad Endings as Britain's Little Wheels Come Off

Proper Dinky Toys, the groundbreaking range from Meccano that had kicked off the die-cast craze in 1933 and whose brand name had become a metaphor for the whole toy car genre, failed to make it into the 1980s by just one month. Its trade brochure for 1980 was already printed and many had been sent out; now most would be tossed into wastepaper baskets in toyshops across the land as proprietors considered what would fill the gap on their shelves.

Several of the vehicles in the 1979 catalogue had never actually materialised, such as what looked like a very promising idea for collectability: model coaches in football-club liveries. Some of those that did arrive on sale were ham-fistedly customised versions of the existing Range Rover, Land Rover, Ford Transit, Chevrolet Corvette and Plymouth Gran Fury … only some twelve years after Hot Wheels first exploited the custom-car theme! Other relics still in production at the end included spacecraft from *UFO*, Gerry Anderson's sci-fi TV show first aired in 1970.

The demise of Meccano Ltd had been a slow-motion disaster. As far back as 1976, an internal report had found that production costs for Dinky Toys were 20 per cent more than at rivals Mettoy and Lesney and over the following four years the company accumulated losses of £4 million, with management evidently unable to cut overheads or boost efficiency. At the same time, unions such as the National Union of General and Municipal Workers had made employees so militant that working practices were all but impossible to update and, on top of that, absenteeism and pilfering from the plant was rife; a few Dinkies, rejects or not, were reportedly regarded as a 'perk' of the job – shiny toys that perhaps Liverpool kids might not have had a chance to enjoy otherwise.

Assembly-line staff – mostly ladies – not surprisingly enjoyed the work, the camaraderie and the continuity of relationships. Reading online forum discussions between former employees, and their children, there was a thriving social aspect to working life in the ageing plant between the 'mums', with raffles, bingo and get-togethers around mail-order catalogues and the selling of Avon cosmetics between long-time friends Tricia, Doris, Edna, Evelyn, Violet and the rest.

At the end of 1978, rumours of the company's deep malaise were swirling at the annual Meccano Exhibition, held in Darlington, County Durham, and the company's marketing manager Bryan Farrar was moved to issue a strong denial in the January 1979 edition of the venerable *Meccano Magazine*.

'Everyone here is fully committed to building Meccano back to its former glory,' he wrote.' Over the last few years Airfix Industries and the Government have invested a very substantial sum of money in Meccano Limited, mainly because they believe in its future.' Without mentioning Dinky by name, he went on to boast of the company's TV advertising campaign, export push, and a focus on producing Meccano sets rather than components. 'I think you will appreciate that this level of activity and investment would not even be considered if the future of the Meccano product was in doubt.' Could there have been any significance, however, in the departure of long-time magazine editor Chris Jelley, announced in the very same issue? After eighteen years with the company, he told readers he was leaving to become press officer at National Girobank, based not far away in Bootle, Merseyside. For the last seven years he had been *Meccano Magazine* editor, but now his bubbly descriptions of each month's new Dinky Toys releases would be no more.

When the crunch finally came, it was sudden and astonishingly brutal. On Friday, 30 November 1979, Airfix decided abruptly to shut Binns Road, giving the 900 employees less than an hour's notice that their jobs had ceased to exist, that they should get their coats and go home. Staff left the building in tears, the family atmosphere they'd come to rely on instantly shattered.

There then followed an ugly and infamously bitter worker occupation of the factory that lasted a scarcely believable four months. 'The only course of action open to the workforce was to take control of the situation, to escort the management out of the factory and occupy it,' recalled former union official John Lynch in the *Liverpool Echo* in 2019. 'The shop stewards got together and made the decision that we weren't leaving.' Initially, some of the 100 staff taking part actually continued working in the vain hope that the business might be revived. They were sustained by donations to a fighting fund, looked after by a hardship committee, which provided food – including a hot Christmas lunch on 25 December. During this period, though, a lot of the fixtures and company records are said to have been vandalised and trashed. Workers kept a vigil at the gates throughout that dark, cold winter and the factory's frontage, once maintained immaculately, was daubed with graffiti by supporters, including the slogans 'Bomb Airfix!' and 'They fixed us, now we'll fix them'. Finally, however, the dramatic, emotionally charged protest came to nothing. Airfix succeeded in gaining an eviction order and had the downcast former employees thrown out by bailiffs, assisted by police, in April 1980. The lasting pain of this episode, infused with recrimination, anger and blame, means there are almost no written accounts from insiders. It's hard to imagine a more depressing ending to such a once-proud enterprise.

I can recall clearly that stocks of Dinky Toys lingered in shops for months and even years afterwards, their fading blue, red and yellow packaging barely helping to shift them, until they were marked down and put in bargain bins. And, of course, Airfix Industries itself went bust in 1981, weighed down principally by debts incurred by Meccano Ltd itself (more than £700,000 in 1978 alone), so the closure of Binns Road did almost nothing to save its ultimate owner anyway.

As all this was happening in Liverpool, down in east London Lesney was fighting for its life. The odds were stacked against it. Buying AMT had been a costly mistake for which the company was now saddled with crippling interest repayments, while the strong pound was hobbling the vital export sales that underpinned its turnover. A nod had been given to the high labour costs in the UK by contracting out a range of Walt Disney-licensed vehicles to Universal Group in Hong Kong, but there was still a substantial UK payroll of 3,500. The £10 million profit highpoint made in 1976 had tanked to a loss of £3.5 million in 1980 and Lesney's banks were getting understandably alarmed. It appointed an executive chairman and managed to persuade Jack Odell out of retirement to return as co-chairman to oversee the die-cast and engineering activities. Belt tightening and cost reduction began at the same time as a search for a takeover white knight.

Yet, despite a huge improvement in productivity, by the summer of 1981 the banks refused any more credit and on 11 June the firm was declared insolvent with its shares suspended at 11p each. This valued Lesney at a paltry £3 million. 'Once the cashflow starts to slow, you find that the banks are not willing to help you through a difficult period,' Leslie Smith reflected at a 2004 exhibition devoted to Lesney's importance to the community of Hackney, east London. 'It was a sad story but there are many good memories.'

The receivers put Lesney's assets into a holding company that it named Matchbox Toys Ltd and in June 1982 began negotiations to sell it. Observers anticipated the buyer would be a big American toy firm like Hasbro or French die-cast car rival Majorette. The surprise suitor, however, was Universal Group of Hong Kong, headed by the dynamic David C.W. Yeh. This was the man who, with that Disney series, had shown Smith and Odell that Far Eastern quality could match that of their London plants. He'd even produced a couple of sample extra designs for the 1–75 Series, numbered 76 and 77 and never actually put on sale, to demonstrate that Universal engineering was also first-rate. The two Lesney mainstays had little choice but to reluctantly concede it was now simply too expensive to mass produce in the UK. Yeh had been involved in 'offshoring' production ever since the 1950s when he worked with US toy-making legend Louis Marx in transferring manufacture out of the US and into low-cost Hong Kong. But Yeh also pointed out the parochial nature of many of the Matchbox vehicles, claiming there weren't enough American vehicles to satisfy the US market. That was not entirely true or fair; the Matchbox catalogue had been peppered with US-origin cars and trucks since the late 1950s, and one of the key things that made the toys so enticing was the diversity of the subject matter ... no matter how unfamiliar little Jimmy in Ohio might have found a Bedford TM or a Ford Cortina 1600 GL.

Now Yeh got the remains of the company for £16.5 million, after having bargained hard to take just a single manufacturing plant, the Rochford one. The receivers, showing unusual concern for the welfare of the workers who would be discarded, tried to insist that Yeh take both the main plants, but the Midland Bank disagreed and insisted Yeh's offer was grabbed before it was withdrawn. So the vast Lee Conservancy Road factory – so proudly venerated in Lesney's 1970 book *Mike and the Modelmakers* – would henceforth lie derelict, a ghostly concrete monolith, until it was demolished in 2010. The deal was sealed on 24 September 1982 and Yeh later said: 'It was the bargain of the century. I didn't realise just how valuable a trade name it was.' The UK head office was now at Burleigh House in Enfield, Middlesex, but Universal Matchbox had its main HQ in New York.

The Chingford R&D facility closed and the staff moved to Enfield in 1983 too, but Universal shifted production of the iconic 1–75 series and Super Kings to Macau, on the Chinese coast, where manufacture and exports began in May 1983. In the process, all mention of 'Lesney Products' was routinely expunged from bases and castings. Departing for Macau with them went the attractive and successful new Convoy range. These had their roots in an articulated truck design introduced in the 'Twin Pack' series in 1979 that recalled the Accessory Pack and Major Pack lorries of the 1950s in that its length and scale complemented many of the 1–75 range constituents. It was based on the American Peterbilt tractor unit with its chunky, locomotive-like nose but the lack of flair to its finish, with a drab black plastic base and grille, caused Paccar, the Peterbilt trademark owner, to refuse permission for it to be branded as such, and Lesney instead stamped it with the anonymous 'Long Haul' underneath. The Long Hauls proved popular anyway, and so the company embarked on a rethink. A new Peterbilt with a sparkling chrome-effect plastic grille and bumper and a gleaming new wheel design was created in two versions, together with a more slab-fronted Kenworth, while a range of plastic-bodied trailers were created alongside that opened up an entirely new Matchbox collecting avenue, and opportunities for liveried editions, that would prosper throughout the 1980s. The company also took the cheeky opportunity to grab the Convoy name that now lay unclaimed after the sorry demise of Dinky Toys and its eponymous model truck line-up. This time Paccar approved and allowed its marque names to be used. For a year, they were produced in London before joining the exodus of British-made toys. They may not be especially realistic but the Convoys (later augmented by Scania, Daf and Ford Aeromax cabs) have a lovely character and an almost infinite variety once limited editions and gift sets are included.

A few examples of a cheapened 1–75-style range called Super GT were made in England, but from 1984 the only Matchbox products to be wholly British-sourced were Models of Yesteryear, produced at Rochford by a staff reduced to 1,000 (when it once had 2,000 workers and had been the town's largest employer), who now also looked after warehousing. And even that operation was closed in 1987, ending British production forever. The one bright spot as far as the UK was concerned was that Lesney's industrial die-casting division survived; it was sold off and remains very much in business today as Lesney Industries.

Corgi entered the 1980s seemingly, to its loyal customers, as buoyant as ever. In 1980 the constant stream of new releases continued as was customary. This included sets and very imaginative individual models such as a mobile Dolphinarium and an HCB Angus Firestreak fire engine with electronic siren and flashing light. Indeed, this one led on to a new range of models under the Corgitronics sub-brand, such as a roadworks Compressor with an authentic concrete drilling noise, a remote control Leyland Roadtrain articulated truck, and a Ford Gran Torino-based Roadhog with a two-tone horn. There was even the Roadshow van containing a real transistor radio. Corgi was on the pace with the latest cars, too, issuing its Ford Escort and Austin Mini Metro almost simultaneously with the debuts of the real things. Throughout 1982 and '83, such relationships continued, with very timely arrivals for Corgi's Ford Sierra, Triumph Acclaim and MG Maestro.

In 1980, the Junior part of Corgi's small toy-vehicle range was axed and they all became plain Corgis.

For a short while, too, full-size and pint-size Corgi versions of the same subject, like the Renault 5 Turbo and Ford Capri MkIII, the James Bond vehicles, and the Volkswagen Polo came together in some irresistible little-and-large sets; the small one to take to school and the big one to admire on your bedroom shelf, maybe.

In 1982 came a brand-new range of 1:36 Classics that now fixated on cars of the 1950s for inspiration including the Mercedes-Benz 300SL, MG TF and Jaguar XK120. In 1983 the Corgimatics line arrived, with concealed buttons that tipped up the skip on the back of the Mercedes-Benz Loadlugger and the tipper body of a Ford Transit. Remarkably, van Cleemput was still at the creative controls as Corgi chased more innovations to stave off the threat from electronic toys. Mettoy itself had entered the digital age with its Dragon, a highly innovative 'personal computer' aimed at the children's market. The company was anxious not to miss a promising new trend ever since it snubbed the chance to make Action Man on the basis that boys would never take to dolls ...

All the time, though, the storm clouds were gathering above Northampton and Swansea. Sales in 1980 fell by a disastrous 3.5 million units to 4.8 million annually (plus Junior-sized vehicles) and Mettoy recorded a loss of £3.5 million. The financial drain raged for the next three years, even as the workforce shrank to about 900 (Mettoy's long-ago peak was 6,000) and all remaining activities in Northampton apart from the engineering department either closed down or transferred to Swansea. Managing director Peter Katz was asked to resign, refused, and so was fired in 1982, but in October 1983 the receivers were called in. The Mettoy era had ended and Corgi seemed to be just another busted flush of Britain's once-vibrant manufacturing sector. As Peter Katz recalled in a 2012 interview:

Mettoy was the last to go, largely because of the Dragon computer. In the years 1979-81, following Margaret Thatcher's arrival, you had I think 25 per cent inflation, which doesn't help if you're exporting. Usually if you have that sort of inflation, your currency will devalue, but it didn't devalue because we hit North Sea oil at the same time. So you had the miserable business of cutting back, cutting back, cutting back to try and maintain your viability. But there's just so far you can go. Toys are pretty simple products but you couldn't automate assembly, for instance, because each little car would have needed quite different fittings. What we didn't do is transfer a big chunk of our production to the Far East and close a factory. That would have been an abomination to the board at that time. I think we, me included, were much more in the ethos of manufacturing.

13

Destiny for the Greatest Names in Die-casts

The foundries, die-casting machines and clattering assembly lines at Meccano's Binns Road factory fell silent in November 1979, never to restart. The 245,000sq. ft site was auctioned off and by the end of 1980 most of the rambling buildings had been razed to the ground.

There were still plenty of unsold stocks of Dinky Toys vehicles in shops up and down the UK. Perhaps some retailers wondered what to do with it all after such a high-profile company had crashed out of business so publicly, and no doubt the pricing gun and sheets of Day-Glo half-price stickers were on hand to start the clearance. However, there were still signs of life, and the Dinky marketing department, now housed at Airfix's premises in Wandsworth, south-west London, had a plan in place by the end of 1980. In fact, they brazenly declared it 'the largest new product schedule ever undertaken by the company'.

The first move was to introduce a range of twenty-four Matchbox Superfast-sized toys at the pocket-money price of 59p each – after Dublo Dinky and Mini-Dinky, the third stab at that potentially lucrative market. This comprised four larger-size motorcycles and twenty cars and vans with garish decals and a few opening features and a handful of those – the Honda Accord, Chevrolet Chevette and Toyota Celica – that could not be found in the established competing ranges from Matchbox or Corgi. The design and build owed nothing to the UK. They were merely a selection of the Universal range made in Hong Kong and sold elsewhere under the Kidco brand. Rebadged as Dinkys, they were sold in blister packs and the quality was well below the par of rivals; they were both cheap and nasty.

Soon after these, another range was introduced. These six European cars, which included the Citroën 2CV, Peugeot 504, Alfa Romeo GTV and BMW 530, as well as models of the newly introduced Fiat Strada and Citroën Visa, were all to a constant 1:43 scale, and the £1.50 price undercut the cheapest cars in the Corgi and Matchbox Super Kings line-ups. Once again, though, they were simply rebranded imports, made by Solido in France and near-identical to its budget Cougar range. There was even more to come as Dinky brand executives looked for a contractor to put some of the former Liverpool Dinky products back into production, which included many of the space-themed toys (the *Space 1999* Eagles still hanging on in there), the Convoy trucks range and the Ford Granada MkII, which had never actually been manufactured before. This could have meant a tie-up with Polistil in Italy because, during the early–mid part of 1979 Dinky tooling for the Volvo 265DL and Zygon War Chariot had been sent to Polistil where some batches were made – the very last-gasp attempt to cut costs at Meccano.

All future plans, though, were voided in January 1981 when Airfix itself called in the receivers – yet another spent force in British toy-making. In the gloomy fire sale of its assets that followed, the Dinky Toys brand name was acquired by US conglomerate General Mills (sugary cereals and toys do have a certain synergy in being targeted squarely at children) via its UK subsidiary Palitoy. But nothing was done at all with Dinky and, in 1987, General Mills decided to get out of toys altogether and sold the Dinky brand to Universal Matchbox. In 1989, Dinky was back with a range of made-in-China collectors' models of mostly 1950s and '60s classic cars in The Dinky Collection. They were accurate and well presented in clear plastic cases to ride the then fervent classic car boom, but they lasted only a year or two. Some of the subject matter, including the Triumph Stag, Ford V8 Pilot, MGB GT, Triumph TR4A and Tucker Torpedo, showed careful thought as to what would be either popular or novel. They were intended primarily as display pieces.

Via another change of ownership with Tyco, Dinky now belongs to Mattel – the old nemesis of the British toy car industry – and in 2015 the name was licensed out to collectables company Atlas Editions, which has issued several copies of classic 1950s and '60s British Dinky originals such as the Triumph TR2 and Austin Mini Countryman. Serious collectors aren't very interested; as ready-made replica 'collectables', they'll never amount to anything in terms of investment value.

David Yeh's Universal Group certainly did get a bargain when it bought the unpromising remains of Lesney. 'The Hong Kong Bank funded it all,' he said of the £16.5 million purchase price. 'I didn't pay one cent.' Once manufacture had been shifted largely to Macau, sales rose rapidly from $82 million in 1981 to

$240 million by 1985. With 45 per cent of these in the USA and 25 per cent in Europe, Matchbox was still one of the world's leading die-cast toy car brands. Capitalising on this, Yeh took his Universal Matchbox Group to the US stock market in April 1986 – the first Hong Kong company ever to make such a move. The listing was four times oversubscribed, thanks to the resonance of the Matchbox name. Six years later, the company was taken over by rival toy firm Tyco for $106.73 million, which handed Yeh a more than ample return on his risk, and he left the toy industry to become a property tycoon in China.

Tyco, in turn, was absorbed by Mattel and the Matchbox name is still going strong as a parallel line to Hot Wheels, its roots in a run-down pub basement in east London unknown to just about everyone tossing a Matchbox car into their shopping baskets in hypermarkets all over the world for a few pence, cents, yen or Chinese jiao. The Yesteryear range had seemed to have life in it beyond 1987 and the end of British manufacture but the 'Made In China' era killed the magic. Still, with an estimated 100,000 serious collectors as late as 1993, a revamp was attempted under the 'Matchbox Collectables' banner. The first issue was a Citroën H Van and the line eked out whatever enthusiasm remained for a few more half-hearted years.

Amazingly, Corgi too was to fall under Mattel ownership, making this a depressing hat-trick of acquisitions by the very company that had triggered the drawn-out demise of all three of the great British names. Mettoy was in limbo for six months between October 1983 and March 1984 while the administrators pondered on what to do with the insolvent business. In the end, they went for a buyout by the existing management, led by Mike Rosser, which had gathered together funding from a consortium including Electra Investment Trust, the Welsh Office, Investors In Industry PLC and Lloyds Bank. Clearly, the financial community still thought there was plenty of mileage left in die-cast vehicles. Corgi Toys Ltd was formed on 29 March, with everything based in Swansea, and the new firm returned confidently to making die-cast vehicles. There were some new releases and plenty of older castings repainted and repackaged, but the company concentrated on two key areas. Firstly, securing bulk orders for cars and trucks that were used for promotional purposes. There were tie-ups with confectionery company Cadbury, oil firm Castrol and petrol retailer BP, this last one such a commercially important and massive undertaking that several older Junior-type models including the Rover 3500 SD1, Vauxhall Nova and Jaguar XJS were re-engineered with die-cast baseplates in place of plastic for an added feeling of quality to please the partner. And secondly, the market for collectibles was chased with vigour, resulting in a huge number of vintage and classic commercial vehicles being issued in a vast array of different liveries and limited editions.

In the USA, sales had been in alarming freefall for some time, now squeezed out by more prominent players. Former Reeves International executive Gene McKeown explained:

> Corgi in truth was not that well known in this country. Matchbox had a name and Hot Wheels had a name, so you had two good die-cast brands everybody knew. Because they were big they had advertising programmes, and they did millions and millions of dollars' worth of business. When we had Corgi we

might do $50–75m a year in business but that wasn't enough to generate money for advertising. Then the independent toy and hobby stores didn't want to carry Hot Wheels or Matchbox because they're competing with the hypermarkets. Our retailers were saying they needed something unique and different and that they can make a profit on because the big guys work on very small margins that a small independent retailer can't.

With all these factors in play, McKeown came up with a different approach to aim Corgi straight at the collector. He came to the UK again and persuaded the new management to consider his scheme for limited editions aimed at market niches. As McKeown had himself been a volunteer firefighter with his local Fire Department for forty-two years, he well understood the loyalty, keenness and mentality of the 2.5 million US firefighters. Why not, he suggested, dig out the tooling for the classic Corgi Toys American LaFrance fire engine last used in the 1970s, and make a limited edition? Produce it in Hong Kong, sell it for a high price, and it should be a hit:

They were able to recover that tooling and get it over to Hong Kong. I went to the owner of Reeves and I said I have enough confidence in this that we can do 5,000 pieces. Apart from baseball caps and T-shirts there's nothing for them [firefighters] to buy as a collectible. They were all gone in two months at $40 each. That's a long way from a million Bond Aston Martins but it all made sense.

More highly targeted limited editions along the same lines followed, including a pumper truck adapted from the LaFrance, and then the focus turned to buses, beginning with one for the Hong Kong market itself. Vintage- and contemporary-liveried Greyhound and Peter Pan buses for the US market followed, proving to be sell-outs in carefully focused local areas:

> When the production moved to China, the quality came up. The production quality coming out of Swansea for the previous three or four years of operation was terrible – the product really had dropped. Corgi at that point, just like many other companies, did not have the headache of running a factory. It was creating a sales company. Of course, this was all before the internet, and our sales force was only selling to independent toy stores and hobby shops, but it worked. I think in 1985–86 Reeves represented 55 per cent of their total export business with this. Our quirky business became bigger and bigger.

Events, however, were to once again make an impact on the resurrected Corgi Toys Ltd, this time ending its venerable position as a British manufacturer. US toy colossus Mattel made an approach to buy the company, which was trading profitably, and in December 1989 the management accepted its offer. But why would they do this, when their own Hot Wheels brand was the world die-cast leader? Gene McKeown gave some insight:

> Mattel was unhappy with Hot Wheels in the UK and they simply felt if Corgi did it they would have better distribution. The other thing they were after was Corgi Juniors because they had some pretty good tooling and also it could give Mattel an

immediate entree into the United Kingdom market. A few of them then appeared as Hot Wheels in certain territories with Corgi castings. They didn't particularly value the Corgi brand, they didn't need another one, it was just a very routine purchase of a sales network.

That, then, was the sorry end for British manufacture of Corgi vehicles. In January 1991 the Swansea site was vacated and all production staff laid off as manufacture was summarily shipped off to China, never to return. The last of Britain's three historic die-cast toy car plants – industrious hubs that at various times worked round the clock to meet insatiable global demand – had gone for good. (On 16 June 2011, with the old Mettoy buildings being used illegally to store shredded tyres, the premises caught fire and the whole place was virtually torn down in frenzied efforts to control the horrendous blaze – gruelling work that took three weeks and cost a staggering £1.5 million.) The sales department was uprooted to Mattel's HQ in Leicester; Corgi was now a brand, no longer a proud company. Gene McKeown recalled:

I would go back to England and meet with the Mattel and Corgi people. The executive vice-president of Mattel USA came to me and said: 'Gene, I don't understand this business. I can't talk to these British guys. What the hell do they do making 5,000 pieces of a double-decker bus? If we sell 250,000 pieces of an item, it's a failure but Corgi's 5,000 pieces of a bus is a success.' I explained to him it's a collector business, a niche market, but he said they're really not adding anything to the bottom line of Mattel.

It's not surprising, therefore, that Mattel chose to sell Corgi off in 1995, offloading it to its then-current management, who promptly retitled it Corgi Classics to reflect the fact that mere toys were the last thing that its detailed showpiece models could be considered. And yet again, in 2000, another new owner called Zindart, an American collectibles firm, took control. Then, amid much patriotic fanfare, in 2008 the Corgi brand and design rights was acquired by Hornby Hobbies for £7.5 million, in whose British custodianship it remains to this day in a portfolio that includes other familiar and hallowed names including Hornby Trains, Scalextric and Airfix.

British collectors couldn't help but feel dejected at the long, slow death of the country's die-cast toy car industry that fed their passions. However, there was one surprising turn of events that provided a national champion to fight back, and it was all down to that determined old warhorse of the business Jack Odell.

He firmly believed there remained an excellent market for British-made die-cast vehicles, and that it could also turn a good profit. So with his new partner Bert Russell he established a company in Ponder's End, Enfield, north London, in the summer of 1982 and called it Lledo, which was his name spelt backwards but one that he'd also used as a mnemonic device for his wireless call sign from his Second World War service days in the African desert. A large quantity of the Lesney production machinery had already begun to be shipped out to Macau. Odell was intimately familiar with almost all this Matchbox manufacturing plant, for the simple reason that he had designed the overwhelming majority of the equipment himself. Now, aged 62, he made the bold move to buy some of it back, which involved shipping the machinery back to

England and salvaging other bits of kit he needed from the various ex-Lesney plants in north and east London that were being dismantled and sold off. Odell and Russell spent eight months equipping and tooling-up their new factory.

This time, the output would be strictly on Odell's terms. He was going to eschew the high-volume/low-return Matchbox 'miniatures' and any attempts to follow the fads and trends of the ultra-fickle toy market. He chose a return to the roots of his Models of Yesteryear, which were to be called the 'Days Gone' series and would start to build a range of collectable models of the veteran and vintage vehicles Odell liked and was familiar with from his own childhood.

The new product line-up was launched in April 1983, and rather surprisingly the first five were all horse-drawn vehicles to an undisclosed scale smaller than recent Yesteryear issues but larger than the original Yesteryears of the mid 1950s. The horse-drawn tram, milk float, delivery van, omnibus and fire engine were indeed in the spirit of the subject matter Odell had fixated on more than thirty years before. Each one came with a set of plastic figures on a sprue frame to add human interest. Those with an automotive bent were probably left cold until, that is, a Ford Model T van joined them in autumn 1983.

And it was with this issue Lledo took off. Its flat sides were a natural blank canvas for liveries and advertising and the new company began a marketing drive to use the Model T as the basis for what became an enormous number of different promotional models made for specific clients and marketing campaigns, and as corporate gifts. The business model was very different in that it responded to the pull of customer demand rather than a relentless push into the retail market.

Over sixteen years and more than 6 million examples, the Model T helped turned Lledo into a sizeable manufacturer, and a very successful start-up. Therefore, this one basic toy vehicle (different liveries aside) sold more examples than Corgi's talismanic James Bond Aston Martin. It was a highly creditable effort to keep die-cast model manufacturing in Britain where all the famous old names had failed.

From 1984 more and more models were launched – seven that year – the majority of which were 1920s and '30s vans that offered a goodly area of flat side elevation to carry marketing messages, for the Promotionals, and period livery, for the Days Gone fleet. In 1987, Lledo started to issue themed sets (the first commemorated the seventy-fifth anniversary of the Royal Flying Corps) on its own behalf and for paying clients, and before long it had created its very own collecting universe where acquisition of everything it produced presented the sort of challenge that die-cast completists absolutely relish. With a huge variety of unique packaging and irresistible 'certificates of authenticity', people were hooked into this new cult. In 1989, Lledo's evocative set to commemorate the fiftieth anniversary of the Battle of Britain proved to be a watershed in generating £1 million of sales ... a portion of which was donated to the RAF Benevolent Fund.

There was something of a new dawn in 1994 in the Days Gone range: Lledo launched a very finely detailed Morris Minor 1000 Traveller, which broke new ground on three fronts. For one thing, the level of detail, with its fully featured interior, glazing, carefully picked-out wooden frame and chrome bumpers, and the superb proportions, upped the ante considerably. Second, it was to 1:43 scale. And thirdly, it brought the scope of interest surging forward to the booming post-war classic car scene. The excellent reception for the Traveller made it, two years later, the first constituent of Lledo's new Vanguards range of passenger cars and car-derived vans from the 1950s and '60s. Vanguards swiftly evolved with large commercial vehicles at 1:64 scale, and new packaging that returned to the traditional presentation boxes so beloved of serious collectors, and the level of detail in terms of parts like chrome headlights and wing mirrors started to reach levels not seen since the delicate Spot On models of the 1960s although, unlike those beauties, Vanguards were intended as showpieces and the only children who ever went near them tended to be grown-up ones.

By 1998 Lledo employed 300 people in Enfield making 6 million very attractive items annually. Regrettably, this happy situation was not to last much longer. Jack Odell was knocking on and, perhaps aware of his own mortality and with no one in the family to take his mantle as the totem of British die-casting, in 1996 he sold a controlling stake in Lledo to another company. When that went bankrupt in 1999, Lledo then became simply an asset in receivership. A buyer was sought and that buyer turned out to be Corgi. On 16 November Corgi was able to purchase the bits of Lledo PLC it wanted for £1.95 million. That included the brand name, designs and tooling but not the Enfield factory, which was shut down as production was immediately relocated to China. Corgi at least kept the Lledo, Days Gone and Vanguards trademarks going for five years, but in 2004 even these were dropped as any surviving models were folded into the Corgi Classics range.

Britain's toy car wars, then, were finally finished. It had all gone. But at least Jack Odell was to enjoy eleven years in retirement (on top of his comfortable seven-year sabbatical in the 1970s) as a living legend among collectors before he died in Barnet, Hertfordshire, in July 2007, after battling Parkinson's disease, just two years after his erstwhile partner Leslie Smith also departed for that great toy factory in the sky. Odell left an estate of £8.6 million to his widow. And that, really, was the sole dividend for this country of all the efforts relayed on the previous pages herewith. We do, however, have millions upon millions of surviving toy cars to keep the glow of it all rosy for many generations to come.

14

Celebrity Status: Icons Inspired by the Big and Small Screen

The first brush with the powerful influence of the moving image experienced by any of Britain's three die-cast giants had nothing to do with cars. It had nothing, indeed, to do with vehicles whatsoever.

Lesney's sales partner, Richard Kohnstam, had spotted the potential of a tie-in with the popular children's puppet show on the flickering, black-and-white BBC television service called *Muffin the Mule*. He put in an order to Lesney for a die-cast metal replica, with the ten-piece, articulated puppet operated by four strings, after seeking permission from the presenter Annette Mills and the designer Ann Hogarth. There's no mention of Lesney at all on the puppet or the box. It sold very briskly and only reached the end of its shelf life in 1955 when Mills, sister of actor John Mills, died unexpectedly. Jack Odell and Leslie Smith didn't pursue any more entertainment tie-ups, presumably because of the runaway success of their own Matchbox and Models of Yesteryear ranges.

The tag 'as seen on TV' began to exert an ever greater pull for marketing just about anything in the late 1950s and early '60s. Die-cast minnow Budgie briefly jumped on the passing bandwagon with a neat die-cast model of the Supercar from the eponymous Gerry Anderson puppet show. However, it was Corgi that truly stumbled upon the gold mine awaiting a successful collaboration between model cars and motion media. In his role of regional sales manager for Scandinavia, Mettoy's Peter Katz had one product in his arsenal that he could really make inroads with: Corgi's Volvo P1800 that the company had launched in July 1962. It was a straightforward model with its spring suspension, moulded interior with steering wheel, and jewelled headlights.

Katz recalled: 'Going to Sweden quite a lot, talking to customers, I asked: can we do anything special? One of my wholesale customers said: "Well, you know, a very popular TV programme in Sweden is *The Saint* and he drives a Volvo P1800; why can't you sell that as the Saint's Volvo?"'

Katz took the idea back to Northampton and van Cleemput regarded the transformation as 'a natural'. In what was one of his easiest design projects ever, he put the Saint 'stickman' emblem on the bonnet as a decal and installed a plastic driver as an approximation of actor Roger Moore's Simon Templar. Together with new box artwork and eager approval from ATV (although, according to van Cleemput, no royalties paid for a licence), the effect in 1965 was electric. The standard Corgi Volvo sold 315,000 examples in three years but the Saint edition shifted 321,000 in the nine months of its first year on sale alone, going on to sell 1.2 million examples in total. 'I think I had one minor coup [with it] in my export sales days,' recalled Katz. 'It sold extraordinarily well.'

What Corgi did next, though, utterly eclipsed the Saint's Volvo. Its James Bond Aston Martin DB5 took the toy market by storm in October 1965, creating pandemonium in stores across the country and fuelling a media-led frenzy as parents scrabbled to grab one in the Christmas stampede.

The first inkling that there was something worthwhile for Corgi in the movie exploits of James Bond came on 14 September 1964, three days before the London premiere of *Goldfinger*. Four pictures of the Aston Martin DB5 that would appear with Sean Connery in the film appeared in the local newspaper in Northampton. It was plonked on the desk of Corgi brand supremo Howard Fairbairn 'as something we should get on to quickly', recalled van Cleemput. Fairbairn's reaction was negative, feeling that it would be too complicated. However, after three months had ticked by and the global success of the film snowballed, Fairbairn changed his mind abruptly and suddenly everything else Corgi-related was halted as the project was rushed through at breakneck pace.

'They suddenly said yes and all hell broke loose, everything else was put on hold,' recalled Corgi designer Tim Richards in a 2017 interview with the author. 'The film was already out and they'd really dragged their heels on it. We already had an Aston Martin DB4 in the Corgi range and I had to modify that into a DB5 and make a resin mould mega-quick. It was a botch job, to be honest.' There was simply no time for the usual modelling process involving painstaking drawings, beautiful 1:12-scale masters in hardwood, and accurate reduction by pantograph to produce the moulds for die-casting.

In emergency meetings, it was decided that three main features of the film car must be included: the ejector seat with a flip-up roof panel through which to dispose of 007's Korean adversary; a pop-up bulletproof rear screen; and front bumper overriders that flipped out to reveal concealed machine guns.

The task of engineering the internal mechanisms fell to John Marshall, initial project leader, as he also recalled in a 2017 interview with the author:

> No one was sure how it would work but I got the car functioning in a week. That's when they gave it the go-ahead – I kicked it off, really. I cut the aperture in a DB4's roof and figured out how to do the ejector seat. Plastic slides would wear out in a week, so I made an arm and bearing across the back-seat moulding, on two pins in a bearing housing. It was frictionless, and incorporated a butterfly spring in the release.

To open the roof and trigger the ejector seat to jettison the villain, John positioned a tiny release button under the Aston's door sill. There was a similar control for the concealed machine guns that deployed as the bumper overriders shot forward, while the pop-up bullet shield in the boot was activated by pushing in the exhaust pipes – all John's work. There are eight patents relating to the car's design work with his name on, filed under the Mettoy umbrella.

Some 30,000 man hours were poured into the design of the toy, with the tooling for the twenty-eight separate components costing £45,000. Although the on-screen car was silver, when this was replicated for the Corgi version it simply looked like it was unpainted, so the company decided to spray the car gold for production.

Corgi management had to be convinced the miniaturised mechanisms were robust enough to withstand being played with relentlessly. And it all had to work properly. Just as real car manufacturers tested their parts rigorously before getting the production line rolling, Mettoy now did the same. John was flat out once again:

> We needed to see how the car would stand up, and so I built my own test rig. I settled the car tightly in a kind of nest, and then positioned just above the roof an electric motor from a shop window display turntable, which rotated an arm at about 4rpm. I made the arm like a simulated aircraft undercarriage leg. As it came round, it brushed the sprung roof to close it, which re-set it, and then a slight extension of the shaft hit a spring-steel strip that flicked the button. The cycle lasted 15sec and it ran 20,410 times until it failed.

The car came packaged in a superb box with display stand and 'secret operating instructions', a lapel sticker and a spare baddie. The fact that it hit the shops in October 1965, almost a year after *Goldfinger* had opened at cinemas, mattered not one bit as far as demand was concerned.

Corgi was deluged with orders and the Swansea workforce worked around the clock to satisfy demand. The Welsh ladies couldn't assemble the DB5s fast enough, as additional 35ft-long articulated trucks waited outside the plant to whisk the Astons down to outlets such as Hamleys and John Lewis in London's West End.

Newspaper reporters descended on the stores' toy departments to report on the clamour. One shop assistant at John Lewis told *The Daily Mirror*: 'Don't mention that 007 car to me! We get about 2,000 inquiries a week for it. And we have not had any for ages.'

Mettoy's advertising manager Bill Baxter also sounded overwhelmed when he told the *Mirror*: 'We never in our wildest dreams expected such phenomenal sales. We have now doubled production but many children will not be able to get one for Christmas.' When supplies did reach shops, they sold out in hours. People would go to extreme lengths to lay their hands on this toy car. Arthur Katz even had his bank manager at National Westminster Bank on the phone, pleading to get hold of four of them for his grandchildren.

Through a Herculean effort, Mettoy's Swansea plant managed to produce more than 750,000 of the cars up to the end of 1965. It really was all hands on deck; Corgi engineer John Marshall remembers his mother and sister both pitching in in Northampton, joining the cottage industry of outworkers who packed the spare villains, stickers and instruction leaflets inside the secret compartment of the inner packing piece of the box. Lorryloads of complete packaging would then make the four-hour road trip to Swansea for the Astons to be packed.

It won the inaugural Toy of the Year award from the National Association of Toy Retailers in 1966. The original version went on to sell more than 4 million examples. Then an upgraded car fitted with tyre slashers, swivelling number plates and painted in authentic silver, launched in 1968, added 1.2 million more and subsequently, in 1978, a completely retooled 1:36-scale edition put on 1 million-plus more. Millions more were sold in the smaller Husky/Corgi Junior size, somewhat incongruously here, as it was actually an Aston Martin DB6 sporting that lethal ejector seat.

From that point on, Corgi went all out to partner successful TV shows and films. For 1966, it was relatively simple to adapt the existing Corgi Classics Bentley 3-litre and the Corgi Lotus Elan into a gift set themed around *The Avengers*, and this was a healthy seller indeed. However, Corgi's Batmobile issued that year in October eclipsed even the first Bond Aston for sales of a single model, with some 5 million eventually sold. It fired rockets and had a chain slasher at the front, as well as featuring plastic figures of Batman and Robin. From the moment Howard Fairbairn glimpsed the photographs of the TV star car in New York to the first models reaching stockists took just nine months. As John Marshall had become, almost overnight, Corgi's resident gimmicks guru, his skills were once again urgently called upon in 1966:

> On this one, some of the ideas came from our bosses and some were me. The sequential triple-firing rockets I achieved using a piece of spring steel with three 'fingers' on it, and you operated it by flicking a thumbwheel. The chain-cutter at the front was fairly straightforward. But the thing I thought up was the plastic jet flame pulsing in and out at the back, driven by the turning of the back axle, simply because it would be fun!

The fact that *Batman* was a staple of Saturday-morning children's TV throughout the 1970s sustained Corgi Batmobile sales for year after year.

Corgi's Thrushbuster from the US TV series *The Man From U.N.C.L.E.* was a little easier to produce, even with its two gun-firing occupants and bullet-hole-sprayed windscreen. Corgi co-opted the casting from its standard Oldsmobile Super 88 saloon car, although no such vehicle ever appeared on screen (that was a gull-wing Piranha coupe specially built for the show, which Corgi did in fact model, but only as a smaller Husky). The Thrushbuster was the subject of some tricky internal controversy at Mettoy. For this, John Marshall had adapted one of his imaginative ideas for a getaway car, with the driver figure firing a gun from an open window:

> I'd been playing with a pressed metal 'clicker' on my desk for a while, and I suddenly realised it could sound like a gunshot. But then my boss, Howard Fairbairn, felt it wouldn't be morally right to present this as a man firing *at* a police car, which was my original concept. So we turned it round: we used it with figures of Napoleon Solo and Illya Kuryakin in our Man From U.N.C.L.E. car – with bullet holes in their windscreen, they were firing *forwards* in defence! It was a huge success.

US importer Werner Fleischmann, head of Reeves International, saw his sales explode thanks to the James Bond Aston Martin DB5, according to his son in a 2021 interview:

> My father used to carry around a James Bond Aston Martin DB5 in his pocket – the most successful product he ever had of Corgi's. When he travelled all over the world, especially to the New Year trade fairs in Europe – at the beginning of each year my mother never saw much of him – he would pull it out to show people. I believe the actual Aston Martin used in the film was once displayed at the FAO Schwarz or Neiman Marcus department stores, which obviously was tremendous publicity for him. After that came the Batmobile, which I think exceeded even the Bond car. Corgi was hot. My father's two favourite words were 'quality' and 'fantastic', and that was Corgi.

So lucrative had the Bond Aston and Batmobile been that everyone involved with Corgi Toys were henceforth on a frantic search to find similar subjects to mine. Reeves International executive Gene McKeown recalled how these two electrified his company's sales in the American market:

> The Goldfinger Aston Martin sold almost one million items in the US, and that was a huge amount of business for Mettoy. Then the Batmobile sold even more. I think this made them really pay attention to the American market and also realise that licencing with movies and TV programmes was one area that really had not been addressed by Mattel. That window was still wide open and so we said let's go after Superman and so on. But the James Bond licence really had the legs because it wasn't just one movie but a whole franchise, the biggest before Star Wars. James Bond vehicles, basically, kept Corgi going.

Among the huge successes there were one or two failures, one in particular based on unexpected consequences, as proved by comic book hero The Green Hornet.

'I'm afraid Mettoy truly got stung with that one,' recalled Tim Richards with a chuckle. 'They went ahead with it but the show never appeared on British TV!' John Marshall was tasked with creating 'Black Beauty', the customised Chrysler Imperial used by the masked crime fighter as he graduated from cartoon strip to half-hour TV show. John recalled:

> In the comic it had a drone in the boot, and I spent a week experimenting with this. I thought we could mould it like a propeller, and that when you opened

the boot it would spin and fly upwards. After a week of experimenting, we thought that proved a bit too dangerous so we settled for a spring to flick a spinner out.

Both John's and Tim's talents were required on Corgi's elaborate and clever Chitty Chitty Bang Bang in 1968. It cost £100,000 to design and tool, and sold 776,000 copies in five years. Mettoy bosses were initially flummoxed on a way to release the concealed wings. John quickly devised the solution; a push on the sprung, veteran-style handbrake saw them ping out. A team from Corgi visited the movie set at Pinewood Studios, and Tim recalled an issue with the film prop car itself.

'We had no idea what the bottom of the Chitty Chitty Bang Bang looked like,' he said. 'There was no reference, and it was the usual rush job.' Indeed, no one could be proud of the car's underbelly, as it was an unsightly Ford Zodiac floorplan roughly cut and welded to support the faux-vintage-style bodywork on top. So in one evening, a Corgi sculptor fashioned it to look like the underneath of a boat. 'It was total fiction – dreamt up that night,' said Tim.

At first, all that Corgi's rivals at Dinky Toys could do was watch helplessly from the sidelines. As The Saint's Volvo was selling like hot cakes at toyshops nationwide, its initial response was feeble and slightly desperate. For Christmas 1965 it came up with the bizarre Dinky Beats, a vintage Morris Oxford with a pop group on board clearly modelled on The Beatles, although the musicians playing guitars and a mouth organ were manifestly a mop-topped Fab Three – the multi-coloured car was evidence that clumsy old Meccano wasn't accustomed to negotiating rights with third parties. Which was a shame, really, as both

The Beatles and Dinky Toys were some of Liverpool's best-loved exports. The car was merrily described in the *Meccano Magazine* under the heading 'Mersey Beat In Dinkyland', trying coyly to feed off John, Paul, George and Ringo.

The company certainly did get its act together in the run-up to Christmas 1966. Finally, it had something to take on Corgi's red-hot Batmobile, and something impressive to boot.

Corgi had initially been in the running for the rights to make the Rolls-Royce FAB 1 from Gerry Anderson's smash hit TV sci-fi puppet show *Thunderbirds* in die-cast metal, with first prototypes already begun, when it was announced that Dinky Toys was getting the rights. Corgi's Marcel van Cleemput has always been emphatic that his company never paid for a manufacturing licence – cannily realising any entertainment producer would exalt in the free publicity the toy car generated. So it would be fair to assume Meccano actually opened its wallet for the privilege. As 'made under licence from Century 21 Toys' was cast into the baseplate, there is evidence of contractual obligations and it's recorded that, much as the company urgently needed to push into 'character merchandising' to be competitive, many people internally at Binns Road considered it a risky gamble.

It was a fantastic toy, no question, well proportioned, with a heavyweight feel, and with rockets firing from front and back to mimic the car's TV antics, niftily activated by pressing down on the sprung suspension. The clear plastic roof was retractable for a better view of Lady Penelope on the back seat and her shifty chauffeur Parker at the wheel.

In 'reality', the bright pink six-wheeler originated as a TV studio prop for the making of the ATV small-screen saga of the International Rescue organisation and the Tracy clan – probably the most discussed and best-loved kids' show of all time. Gerry Anderson conceived the car as transport for the Tracys' British-based affiliate Lady Penelope Creighton-Ward yet the task of designing it fell to special-effects supervisor Derek Meddings, who was told to produce a 'Rolls-Royce for the 21st century', which finally gained cautious approval from Rolls itself so that the unique radiator grille could be used. For the thirty-two episodes, two versions of the car were made, a tiny one about 6in long for inclusion in model sets and another – the definitive one – a 7ft-long wooden model for filming in scenes featuring the puppets themselves. The 'FAB 1' name and number plates (featured on the Dinky Toy) matched the International Rescue 'F.A.B.' radio sign-off, in place of 'Roger'.

In the *Meccano Magazine* it was introduced to readers by excitable reporter Chris Jelley. He wrote:

> I have seen so many innovations, gimmicks and new models, and have seen the diecast modelling world as a whole make so many advances that I had come to imagine I had seen just about everything. Nothing in this field, I thought, could surprise me – then along came FAB 1.

The hyperbole was, in this case, fully justified: it was an instant, massive hit, too, an extremely welcome event for Dinky Toys. One contemporary press report quoted a Dinky spokesman: 'No less than 940 orders – some for as many as 6,000 models – came in within 24 hours of retailers seeing it.'

In 1967, Dinky followed the pink Rolls-Royce with an equally popular rendition of the Thunderbird 2 craft and from then on it made fulsome commercial use of its first dibs on each of Anderson's shows right into the mid-1970s. This meant it was exclusively in the die-cast market with vehicles and flying craft from *Captain Scarlet and the Mysterons*, *Joe 90*, *UFO* and *Space 1999*, and most of them had extraordinarily long retail lives. This was certainly the case for the multi-wheel Spectrum Pursuit Vehicle from *Captain Scarlet and the Mysterons*, *UFO*'s cross-country, missile-firing attack vehicle Shado 2, and Joe's Car from *Joe 90*, in which an AA battery lit up the red glow of a turbine engine in this hybrid automobile/aircraft that, naturally enough, was designed by junior superhero Joe's electronics professor father Ian MacLaine in the convoluted narrative of the TV show itself. All three received star billing year after year in the Dinky catalogues of the 1970s, sustained by continual repeats of the shows they hailed from. They are quite simply some of the most joyful playthings Britain has ever produced.

Dinky was eternally grateful to Gerry Anderson for the success his creations brought to the brand. To a considerable degree the *Thunderbirds* and *UFO* issues saved Dinky Toys from prematurely becoming part of British toy-making history, and its immense gratitude was shown when Meccano presented Anderson with a solid sterling-silver model of FAB 1 in 1971. Sometimes, indeed, Dinky could be too hasty to reach out for Gerry Anderson's stardust in the unpredictable children's toy market. It committed to issuing a car called Stripey the Magic Mini based on a cartoon strip for toddlers that Anderson had hoped to turn into a TV series, only for the show never to get made. Meanwhile, Dinky's Gabriel (a cheerfully painted Ford Model T) from a 1969 Anderson show called *The Secret Service* was an instant dead duck on toyshop shelves because the series was axed after being shown in fewer than half the ITV regions. Another Gerry Anderson show, *The Investigator*, only progressed as far as a 1973 pilot after ITV gave it the bird when it came to commissioning an entire series. Unfortunately Dinky executives were so eager to not miss out that they commissioned the tooling for an eight-wheeled car and a 'spy boat' in advance to make dead sure they were ready for the next TV sensation. With no actual show to capitalise on, Dinky was left with vehicles it had to market as the unconnected Armoured Command Car and the bizarre Coastguard Amphibious Missile Launcher in 1976, with no other special claim than they were designed by Britain's answer to Walt Disney. One other bright spot in the 1960s had been Dinky's Mini Moke from the espionage action show *The Prisoner*, another cult programme from ATV.

This chapter necessarily makes little mention of Lesney. It eschewed licence deals and film or TV tie-ins until 1980, when it launched its range of Walt Disney vehicles based loosely on 1–75 castings and made in Hong Kong.

Corgi, however, worked its way through every media property from which a vehicular link could possibly be exploited. From 1967 that included such long-forgotten TV series as *The World of Wooster*, *Daktari*, and *The Hardy Boys* cartoon show. The Monkees' Monkeemobile by Corgi of 1968 sold poorly, with fewer than 100,000 shifted. And nor did The Yellow Submarine, with figures of The Beatles, fare too well in 1969, probably with a very similar quantity sold.

Both Dinky and Corgi mined the market opened up by animated and live-action shows for very young children. Corgi worked with *Noddy*, *Popeye*, *Basil Brush*, *Wacky Races* and *The Magic Roundabout* through the early 1970s. With somewhat less aplomb, Dinky found commercial allies in *The Adventures of Parsley* and *The Enchanted House*. It also produced a model of the futuristic car featured in the *Pink Panther* cartoons and designed by Hollywood custom car legend George Barris. This featured an oversized heavy flywheel in the middle of its plastic body, through which a toothed plastic strip was pulled to propel it speedily across the floor; it had a rubber nosecone for when it inevitably smashed straight into a skirting board, and was entirely an action toy rather than any sort of model.

In 1976, however, Corgi was back with a serious vengeance. It had moved its focus to Saturday night family entertainment and 1:36 scale, and kick-started this new epoch with the Buick Regal from the *Kojak* detective show starring lollipop-sucking Telly Savalas. It was a very big seller and would be followed in 1976 by an even bigger one, the Ford Gran Torino from *Starsky and Hutch*. Then came the Lotus Esprit from the 1977 James Bond film *The Spy Who Loved Me*; Corgi was so much a part of the launch for the film that it was involved a year ahead of release so examples could be presented to VIPs on its opening night, the first being handed to Princess Anne.

Naturally, the Jaguar XJ-S from *Return of the Saint* came in for the Corgi flattery treatment in 1978, and then the toymaker plundered the Marvel superheroes canon for madcap vehicles featuring the characters Captain America and Spider-Man, along with DC Comics' Superman. There were tie-ups with must-see TV as contrasting as *The Muppets* and *The Professionals*, but in 1981 Corgi was saved from an expensive mistake. Its catalogue was printed with tempting details of the space vehicle from a small-screen outing for British comic-strip hero *Dan Dare: Pilot of the Future* and the prototypes were rolling around van Cleemput's ever crowded desk. However, when the show was cancelled before anything had been filmed, Corgi wisely scrapped its model, saving itself an awful lot of embarrassment. Instead, it went for a real-life fairy tale, issuing the Austin Metro in purple, and a special box, to commemorate the nuptials of Prince Charles and Lady Diana Spencer. The 15,000 made were very hard to sell and history has ultimately adjudged it a failure, rather like the royal union itself.

15

Collecting: A Personal View on Survival and Big Money

Ican't tell you how many Dinky, Corgi and Matchbox die-cast vehicles have passed through my ownership. Literally – I've lost count. It must be in the tens of thousands, possibly more than 100,000. And thanks to a knowledge bank I've built up over fifty years, they've served me very well. My original collection, accumulated between the ages of about 10 and 22, I sold for funds as a deposit for a flat. Then, in recent years, I've traded countless examples online. Through some of the leaner years of the post-2008 financial crash as a writer, this certainly helped keep the home fires burning. At anything from a car boot sale to an upmarket auction, I can glance at a job lot in a cardboard box and know instantly what has value and what is trash. And they don't need to be the earliest pre-war Dinky Toys to be worth good money.

Of course, plenty of people also have specialised knowledge and share it on the internet. So if everyone can look things up speedily, you'd think the era of bargain buys – the time when you can start your collection without spending £100 a time on each piece – is long gone. However, that's not the case. I find plenty of gems via online auction sites. And great finds are still surprisingly regular at village fetes and charity shops. The fact is a toy car made in 1981, if in excellent undamaged condition, can often be mistaken for something that's almost brand new by the uninformed; not much has changed, materially, about them in forty years. At the other end of the scale, if you want to invest straight into the cream of these vintage toys, there are specialist auction houses such as Vectis, which attract collections from all over the world, as well as the buy-it-now pieces on eBay that give instant gratification, usually at a premium price.

Collecting British die-casts began in the early 1960s and was galvanised by the publication, in 1966, of Cecil Gibson's seminal work *History of British Dinky Toys, 1934–1964*, which was one of the first books to make sense of and catalogue the entire output of the brand. It's a surprisingly readable, enjoyable little book, with much attention on the pre-war issues that, even then, were becoming hard to find. This applied very much to the 28 Series delivery vans with their varying liveries that for many early collectors were something of an obsession.

Gibson also stoked interest in post-war Dinky Supertoy Foden, Guy and Bedford commercials and their glorious corporate regalia advertising Heinz 57 Varieties or Regent Petrol. The new hobby started as a passion for a few eccentric nostalgists, grown men who didn't want their childhood ever to end, but soon gained legitimacy once London auction houses such as Phillips and Christie's started to include Dinky Toys in their sales of old toy trains.

The cheerful husband-and-wife duo of Mike and Sue Richardson, via research, publishing and promotion, did an immense amount to lay the foundations of what became a fascinating new area of collecting. They even went to the extraordinary lengths of manufacturing and issuing their own Mikansue models, such as a 1:43-scale Jowett Javelin that Dinky had at one time considered for production but then rejected – filling in imaginary gaps in the Binns Road lexicon. It was emblematic of a previously unseen longing for completism ...

In August 1971, in its local church hall, the Portsmouth Static Model Vehicle Club started a new trend when it held the first of a new type of event. They called it a 'Swap Meeting' because they viewed it as an opportunity for the emerging collector community to simply get together and exchange their surplus scale vehicles. It would become the first of the swapmeets that spread nationwide, and indeed across mainland Europe, as dedicated collectors' fairs that were similar to those staged for lovers of militaria, stamps, coins, records and, today, vintage clothing. With the arrival of specialist dealers, though, the quaint notion of kindred spirits doing swapsies quickly vanished, and swapmeets became the very last place you'd be likely to pick up an overlooked gem. By the early to mid-1970s, pre-war Dinky Toys were already changing hands for decent money and the effect spread to the rest of the issues in the spirit of those original ones – that is, without windows or plastic parts but with their original boxes.

My own dad often used to while away winter evenings at his desk, undertaking amateur restorations of very battered Dinkies from his own childhood. It was 1973–74, I was 8 or 9, and I seem to recall that I helped in some way as they were dismantled, repainted (with little tins of Humbrol enamel) and then lovingly reassembled. This activity was possible thanks to North London company Pirate Models. As a sideline to its core business of making white metal kits of buses to enliven OO-gauge railway layouts, it had astutely realised there was enough pent-up demand for Dinky Toys spare parts that it decided to go into the remanufacturing business for them. It started making die-cast replacement radiator grille units for the much-loved 24 and 30 Series cars and 25 Series lorries; these invariably suffered broken bumper ends or headlights when they had been played with in the 1930s and '40s, and these new grilles restored their handsome completeness.

By 1977 Pirate had gained permission from Meccano to reproduce the whole baseplate for 24 Series cars that had fallen apart, at £2.50 each. Replacement plated car radiators were 30p each, un-plated lorry radiators 15p each. To reboot tired examples of the adored 36 Series sports cars there now were offered two replacement headlights and a steering wheel for 15p, sets of brand-new wheels at 30p, plus tyres at 50p. Of course, although it was fun to breathe new life into these old toys, the copy-pattern spares didn't increase their value; only largely undamaged originals began to see steep climbs in their values, and they became increasingly hard to find anywhere.

By the late 1980s, all but the final few years of Dinky's output was regarded as collectable and were now joined by Corgi and Matchbox products of the 1950s and '60s. Spooling forward some thirty years, anything and everything from the three brands is sought after, just as long as it's from their Made in England/Great Britain period. Even then, pre-1980s

Corgi and Dinky vehicles sourced from Hong Kong are sought after (indeed, Made in Hong Kong itself isn't a no-no any more for all kinds of vintage toys, especially brittle plastic cars and lorries from the 1960s made by such spectral copycat companies as Telsalda, Lucky Toys, Marx and Blue Box). In the Matchbox sphere, the short period of manufacture in Macau is one that's also hotted up.

Pretty much everything from the factories in Liverpool, east London and Swansea were mass-market items, usually made in quantities from the hundreds of thousands to the multi-millions. Condition is important, of course, but at these sorts of volumes, you can expect to find almost everything sooner or later in very good to excellent condition.

'Mint' condition would only apply to a model that had either never been played with at all, or was treated with supreme care by its original owner. These, after all, are toys, meant for carefree enjoyment. A varying extent of chipped paintwork is the normal state of discovery. Most boys had many more than one model and when they're casually thrown together in a toy box or drawer, or rummaged through, they will chip further. The mazak alloy can't rust if the paint is damaged (unlike on real cars) but the stony grey metal of the casting will peep through. A car or lorry that's been really roughly played with may have virtually no paint left at all.

Die-cast bodies are tough, designed to stand up to youthful exuberance. Slender screen pillars supporting a roof, though, can break under pressure, although most 1940s and '50s Dinky Toys are so strong they can resist even that. From the moment plastic parts were introduced by Corgi in 1956, though, the opportunities for damage multiplied and many of the most elaborate Corgi and Dinky models of the 1960s have small plastic components that were especially prone to breakage. Plastic windows can be grazed, cracked or broken all too easily.

In adding plastic inserts to simulate glazing in the real car or commercial vehicle, the manufacturers were concluding a process that went right back to the dawn of Dinky Toys' existence. While Meccano construction components allowed for infinite ways to assemble and deconstruct models, die-casts were designed from the start to be 'sealed' units, gaining strength, solidity and lasting play value from the fact that they were all-of-a-piece, the interconnecting nature of the components meaning that the toy would need to be forcibly broken apart to remove, say, a wheel or the rear bodywork of a lorry. Once a machine-flattened spigot, resembling a rivet head, was used to hold the baseplate snugly in place, which in turn located the axles in their correct positions, the sealed-forever nature of these toys was embedded, and it continues to this very day. The consequence for collectors is that the vast majority of die-casts just cannot be invisibly repaired via disassembly although, happily, this also makes fakery extremely difficult to pull off.

As detail and added features increased during the 1960s, many cars and trucks started to come with tiny plastic accessories – everything from suitcases for the boot to loads for farm trailers, and all manner of figures to go on and in the various vehicles. All these tiny items were easy to lose anywhere in a young person's home and a large number will have been sucked up into vacuum cleaner bags or thrown away inadvertently. Some are available as remanufactured replacements but the original 'baddie' from a Corgi

Can't Get Enough of Those Vans ...

Dinky's near-legendary 28 Series vans from the 1930s have long been at the high-value epicentre of the collecting world. Nonetheless, there were gasps at a Christie's auction in London in 1994 when a near-impossible-to-find example carrying the livery of the Bentalls department store, in Kingston-on-Thames, Surrey, sold for £12,650. This was a world record that stood until 2008, when Vectis Auctions sold an even earlier 22 Series van, bearing W.E. Boyce's Highgate, London cycle shop insignia, for £19,975. At the time that set the highest price ever paid for any individual Dinky Toy.

James Bond Aston Martin or the surfboards from the roof of Corgi's Morris Mini Traveller now have a collector value in their role of being able to complete a model satisfyingly.

Spring suspension is another difficult area. If the axles pop out of their positions, then in most cases they can't be put back in place because the vehicle is so indomitably 'sealed' and so can't be dismantled for repairs. The only exceptions to this are models held together with screws, like most of Dinky's post-1970s range, the Lone Star Flyers, and Corgi's mid-1970s Tanks and F1 cars.

When you've been collecting for a few years you become used to the innate frailties of specific models. The windscreen pillars on Dinky Toys' Austin FX3 taxi, for instance, for some reason are much more brittle than on others of its saloon cars of the period. In Lesney's 1950s Matchbox 1–75 series, the drop-down ramp of the Marshall MkVII Horsebox and the twin side doors on the Volkswagen Caravette are frequently missing, victims of the primitive experiments in adding opening features to these small-scale models. And then from the 1960s to the 1980s, a continual quest by Dinky, Corgi and Matchbox to add tiny realistic touches in the intricate form of plastic radio antennas and wing mirrors has led to the inevitable result of broken stubs where these tiny inserts were delicately positioned decades ago by deft assemblers' fingers. They could be snapped by play in the original owner's hands or age-induced brittleness at the merest brush today, but the roof aerial on Corgi's De Tomaso Mangusta or the door mirror on Dinky's Monteverdi 375L are almost always missing even if the rest of the car has survived well-nigh unblemished.

Separate rubber tyres, a lovely little feature on most early Dinky and Corgi toys, are susceptible to perishing. They can harden and then crumble, or else shrink and break up. Modern replacements are widely available and won't detract hugely in value terms. In the post-1969 era triggered by the influence of Hot Wheels, separate tyres became a thing of the past but the concern moved to the axles. Matchbox Superfast axles bend all too easily, meaning wheels will be out of kilter, and they're just about impossible to bend back. With Corgi and Corgi Juniors in the 1970s to early 1980s era, the axles are fine but suspension mountings are weak, so the wheels get pushed up inside the car. Once again, this is impossible to remedy if you want to retain original condition.

Talking of renovation, plenty of people like to refurbish 'play worn' die-casts but these retain nothing like the value of an example in original condition, even if it does have a few honest paint chips. There's a strange code system established among collectors that you might encounter and here's the explanation. Code 1 is any model produced totally by the original manufacturer in its factory; Code 2 is any model whose finish, decals or re-packaging has been added by an outsider to an original but generally with 'official' blessing and before it is retailed as-new; and Code 3 is anything that's been repainted, modified or altered after it's been sold to the public but without manufacturer approval. Basically, the first two are pukka and carry originality value and the last one is nebulous, with the results abhorred by serious collectors.

Starting in 1952, almost everything covered by this book came in individual packaging. Surviving original boxes or packs add enormous value to a model or a set, sometimes tripling or quadrupling its value. Such packaging was never meant to be retained by the consumer, so its survival is fluky, and collectors pay a big premium to get it. It's another arena entirely in terms of surviving condition, especially in the areas of box end flaps and the clear plastic windows in later packages. Some 1960s and '70s Husky, Corgi Junior and Matchbox 1–75 vehicles were sold in blister packs and for these to have survived in sealed condition really is a miracle.

With so much extra value at stake, replica boxes are naturally an area that attracts fakers. You need to be familiar with the feel, texture and even smell of genuinely old cardboard to be sure not to get ripped off.

A new and perplexing area of degeneration in die-cast models relates to ageing plastics. Most

Matchbox 1–75 and King Size models of the 1960s with separate plastic tyres, for example, suffer from shrinking wheel hubs. Other plastic parts can discolour in direct sunlight.

Most of the points above have an impact on values, but they also matter when it comes to the enjoyment of your collection. Most of us collectors want our models to look good, to exude originality, and the less damage they exhibit the better. Having said that, really beautifully original items come with their own set of problems. Handling them is the biggest worry, especially when original decals or small parts become fragile with age, and you have to be especially careful when dusting or cleaning them. Just remember: these things were never really meant to be precious artefacts ... which is precisely why they are so collectable today. In fact, that is the whole reason why Lesney's Models of Yesteryear rarely command high prices. They were meant to be admired on a mantelpiece, out of harm's way, and therefore their survival rate in great condition is usually just too high to make them attractive to many collectors.

The holy grail every long-time collector really dreams of, in contrast, is chancing upon a prototype, a 'pre-production' trial colour scheme or, at the very least, an intriguing factory defect that often reflects human error. Actual proposed mock-ups, dies, casts or tooling are indeed extremely uncommon. The chances of them ever having been allowed to escape the inner sanctums at Binns Road, Harlestone Road or Hackney were slim, and most would have been destroyed or recycled. Examples of cars and trucks from the short pre-production runs to test potential colours and liveries, or maybe even try out alternative wheels or small parts, do very occasionally turn up.

Sometimes these issues in unusual paint schemes or with tiny material differences to the retail version on sale in toyshops escaped in the forms of samples given to distributors or shop owners, designed to encourage them to up their orders. As a collector myself, I never thought I'd be able to discover such unicorns, although I always check what appear to be repaints just in case, and think I have a pretty good mental index about what looks run-of-the-mill and what might just be a 'find'. It was in this spirit of endeavour that in early 2019 I paid the princely sum of £12 for a little job lot of toy cars on the prevailing online auction site, about ten in all including a couple of familiar Matchbox cars in the wrong colours that, I assumed, would just be the repainted detritus that came with the other, better stuff. But when the package arrived, the Superfast BMC Pininfarina 1800 in silver and the Ford pick-up in green really did seem to be those rarest of beasts: trial paintjobs done at the factory, the BMC with prototype wheels and the Ford in a colour that, even among the most experienced of collectors, had never before been encountered and recorded. Specialist auction house Vectis was able to verify they were genuine, and at auction they raised £1,080. It's almost certainly the best return on investment I've ever realised. To a lesser extent, rarities come along all the time because the market is random and there are simply so many varieties of Dinky, Corgi and Matchbox products that the thrill of the search just never, ever wanes.

Little Vehicles, Big Sums

There is a healthy market for almost every item in Matchbox's enormous 1–75 output. Collectors, however, lose all sense of proportion when it comes to examples with factory defects or experimental finishes. In 1999, Vectis Auctions offered a Mercedes-Benz 230SL that was a mainstay of the mid-1960s range. Issued routinely with white bodywork, this example instead sported mid-green paint as it was used in 1968 by Lesney management to assess alternative colours (it was rejected). As a unique one-off, one collector just had to have it, and paid £4,100 for the privilege, setting a world record for a small Matchbox car. Since then, though, the demand for rarities has only intensified, propelled by close-up attention to rare combinations of features. In November 2014, a mint Magirus Deutz crane with an equally unusual cocktail of features went for an incredible £8,160. And nor is the frenzy reserved for very early pieces. Vectis raised £4,560 for a Superfast No. 45a Ford Group 6 racing car, rare and boxed and more than doubling its £2,000 estimate.

Bibliography and Acknowledgements

To bring together and interweave the stories of Dinky, Matchbox and Corgi, I have consulted many sources, all of which give excellent insight into their individual areas of specialism.

Books

Ambridge, Geoffrey S., *The Bumper Book of Lone Star Diecast Models and Toys 1948–88* (Geesam-Fossicker, 2002)

Anon, *The Matchbox Annual* (Purnell, 1979 and 1980)

Brown, Kenneth D., *The British Toy Business* (Hambledon Press, 1996)

Brown, Kenneth D., *Factory of Dreams: A History Of Meccano Ltd* (Crucible Books, 2007)

Gibson, Cecil, *History of British Dinky Toys 1934–1964* (Model Aeronautical Press/Mikansue, 1966)

McGimpsey, Kevin & Orr, Stewart, *Collecting Matchbox Diecast Toys: The First Forty Years* (Major Productions, 1989)

Monks, Sarah, *Toy Town: How a Hong Kong Industry Played a Global Game* (THHK/PPP Company, 2011)

Richardson, Mike and Sue, *Dinky Toys & Modelled Miniatures – The Hornby Companion Series* (New Cavendish Books, 1981)

Sasek, Miroslav, *Mike and the Modelmakers: The Story of How Matchbox Models are Made* (Lesney Products, 1970)

McReavy, Anthony, *The Toy Story: The Life And Times Of Inventor Frank Hornby* (Ebury Press, 2002)

Stonebank, Bruce and Diane, *Matchbox Toys* (Quintet Books/ Apple Press, 1993)

van Cleemput, Marcel R., *The Great Book of Corgi* (New Cavendish Books, 1989)

Whittaker, Nicholas, *Toys Were Us* (Orion Books, 2001)

Articles etc.

'All right, James Bond, <u>you</u> try to get me one ... ' (Brian Boss, *The Daily Mirror*, November 1965)

'Cars for Little People' (H.B. Cottee, *The Motor* magazine, 23 December 1959)

'Dinky Toys, 80th anniversary' (Giles Chapman, Octane magazine, May 2013)

'From James Bond, the car with the longest waiting list this Christmas, and at a mere 9s 11d' (*The Sun*, 24 November 1965)

Katz, Arthur, obituary (Giles Chapman, *The Independent*, 27 July 1999)

Matchbox Collectibles, press release Lawson Clarke Publicity, Andrew Tallis and Stewart Orr), 6 July 1993

Mr Brian Mawdsley, designer/innovator: interview by Juliana Vandergrift (British Toy Making Project, Museum of Childhood, Victoria & Albert Museum, 2011)

Mr Marcel van Cleemput, Chief Designer, Mettoy: interview by Ieuan Hopkins and Sarah Wood' (British Toy Making Project, Museum of Childhood, Victoria & Albert Museum, 2010)

Mr Peter Katz, Former Managing Director, Mettoy: interview by Juliana Vandegrift (British Toy Making Project, Museum of Childhood, Victoria & Albert Museum, 2012)

Seeley, Clint, 'A History of Pre-War Automotive Tootsietoys' (*Model Cars* magazine, 1971); with addenda from Robert Newson (www.robertnewson.co.uk)

Smith, Leslie, obituary (Giles Chapman, *The Independent*, 15 September 2005)

Publications

Meccano Magazine (published from 1916 to 1981 variously by Meccano Ltd, Thomas Skinner & Co (Publishers) Ltd, Model Aeronautical Press Ltd/Model & Allied Publications Ltd and Airfix Ltd)

Online

chezbois.com
dinkyworld.com
industrialhistoryhk.org
lone-star-diecast-bk.com
maronline.org.uk
matchboxclub.com
matchboxmemories.com
motomini.com
vintagebritishdiecasts.co.uk
worldcollectorsnet.com/articles
youtube.com/watch?v=q_N1__ZDttQ

Acknowledgements

These are some of the people whose generosity with information and pictures have made this book possible:

Alan Anderson, the late Colin Baddiel, Martin Buckley, Tracey Butcher at The British Toy & Hobby Association, Peter Chapman, Andrew Duerden at Vauxhall Motors, Anthony Fleischmann, Hollie Gaze at the Swansea Museum, Elizabeth Green at the Hackney Archives, Terry Hardgrave, Louise Harker at Vectis Auctions Ltd, Nick Kisch, John Marshall, Gene McKeown, Tom Miano at Serious Toyz, Rebecca Odell at The Hackney Museum, Tim Richards, Julie Sherriff, Karl Schnelle, Sophia Skalbania at Christie's, Mark Tuvey, David Upton, Rod Ward, Emma Williams at the Swansea Museum, Marian Wright, Vanessa Winstone at the National Brewery Centre, Burton upon Trent.

I owe a major debt of gratitude to Hugo Marsh of leading specialist toy auctioneer Special Auction Services (www.specialauctionservices.com) of Newbury, Berkshire, who read the manuscript and shared much of his invaluable knowledge, assuring factual accuracy and guiding me on interpretations. I am also highly appreciative for, and hugely in awe of, the knowledge of highly respected toy historian Robert Newson, who kindly offered me guidance and insight on several issues after reading the original edition of this book.

All images from the author's collection unless otherwise stated.

Also by Giles Chapman

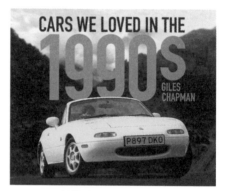

978 0 7509 9318 0

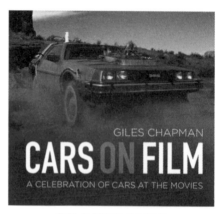

978 0 7509 9400 2